David Riepl

Knowledge-Based Decision Support for Integrated Water Resources Management with an application for Wadi Shueib, Jordan

Knowledge-Based Decision Support for Integrated Water Resources Management with an application for Wadi Shueib, Jordan

by
David Riepl

Dissertation, Karlsruher Institut für Technologie (KIT)
Fakultät für Bauingenieur-, Geo- und Umweltwissenschaften, 2012
Referenten: Prof. Dr. Heinz Hötzl, Prof. Dr. Nico Goldscheider

Impressum

Karlsruher Institut für Technologie (KIT)
KIT Scientific Publishing
Straße am Forum 2
D-76131 Karlsruhe
www.ksp.kit.edu

KIT – Universität des Landes Baden-Württemberg und
nationales Forschungszentrum in der Helmholtz-Gemeinschaft

KIT Scientific Publishing 2013
Print on Demand

ISBN 978-3-7315-0011-7

Knowledge-Based Decision Support for Integrated Water Resources Management with an application for Wadi Shueib, Jordan

Zur Erlangung des akademischen Grades eines

DOKTORS DER NATURWISSENSCHAFTEN (Dr. rer. nat.)

von der Fakultät für

Bauingenieur-, Geo- und Umweltwissenschaften
des Karlsruher Instituts für Technologie (KIT)

genehmigte

DISSERTATION

von

Dipl.-Geoökol. David Riepl
aus Kaufbeuren

Tag der mündlichen Prüfung: 20.12.2012

Referent: Prof. Dr. Heinz Hötzl

Korreferent: Prof. Dr. Nico Goldscheider

2012 Karlsruhe

Abstract

Numerous regions of the world face immense pressure and competition on their natural freshwater resources. The situation appears particularly critical along the Lower Jordan River, where severe physical water scarcity meets very high population growth in a region with a constant potential for immediate conflict. Integrated Water Resources Management (IWRM) is an attractive concept to obtain more sustainable, equitable and efficient water management. But its implementation appears an intricate challenge in itself. On the one hand, the international water sector community strives for ways to translate the theoretical construct of IWRM into operational guidelines for local application. On the other hand, a uniform tenor across disciplines emphasises the need to improve communication and understanding between actors in the IWRM domain, in order to ensure best-possible responses to the upcoming water resources challenges.

Following these insights, the presented thesis takes a two-staged approach to contribute to the contemporary IWRM research. On the one hand it investigates sub-basin-scale IWRM modelling and scenario planning. On the other hand, it develops an approach to collaboratively manage planning and decision making knowledge on the basis of semantic web technologies. The Jordanian Wadi Shueib (~190 km²) is used as exemplary case study for both applications.

Primary water sector challenges in the Wadi Shueib are related to municipal supply, as well as to the management of the resulting waste water return flows. In a comprehensive analysis of the existing data from local institutions and previous studies, holistic monthly water balance time series were constructed. The conceptual system understanding was represented using a water allocation and balancing model (WEAP). National water strategy objectives and action plans were used as normative guideline to craft a set of planning alternatives and performance indicators with relation to the local challenges. Exemplary scenario simulations with a planning horizon of 2025 showed that the current implementation in the study area may fall short of achieving several national objectives. Room for improvement was identified especially in water resources protection and water loss reduction issues. The modelling exercise demonstrates a possibility to provide decision makers with local planning support that is equally based on national IWRM policies as well as on sound science, and thus can be a useful instrument in the progress towards operational IWRM.

In order to investigate the potential of semantic techniques to support knowledge management in the IWRM domain, the modelling study was used as a blueprint. An initial working hypothesis was that ontologies and semantic structures could allow for flexible support of semi-automated IWRM analysis, for example according to established DPSIR (Driver, Pressure, State, Impact, Response) models. Within the work of this thesis, limitations and problems of a generic formal knowledge representation in the IWRM domain within common frameworks were investigated and uncovered. Finally, a conceptual structure and requirements were developed and formalized in an ontology prototype for IWRM planning and decision support knowledge sharing. The concept was implemented as semantic wiki application and opened to the research and stakeholder community of the SMART project. The platform demonstrates a potential to provide solutions for various critical knowledge management requirements. For example to help water sector experts to easily document and formalize their work in a semantic structure, invite colleagues and stakeholders to review and comment, consistently connect their work to previously undertaken studies, and share their insights and work processes with a wider audience over the internet. It is expected that future IWRM initiatives could benefit from comparable approaches, and valuable insights on the subject can be achieved from the pilot study in this thesis.

Kurzfassung

Zahlreiche Regionen der Welt sehen sich mit großen Herausforderungen in der nachhaltigen und gerechten Bewirtschaftung ihrer natürlichen Süßwasservorkommen konfrontiert. Hierzu zählt insbesondere auch die Region des Unteren Jordantales, die zusätzlich mit hoher Wasserarmut, hohem Bevölkerungsdruck und einem hohem Konfliktpotential kämpft. Das Integrierte Wasser Ressourcen Management (IWRM) bietet einen vielversprechenden Ansatz um Wasserbewirtschaftung möglichst nachhaltig, fair und effizient zu gestalten. Allerdings sieht sich dessen Umsetzung wiederum eigenen Herausforderungen gegenüber. Zum einen fehlen bisher meist klare Leitlinien, um das theoretische IWRM-Konstrukt zur praktischen Anwendungen in konkreten Planungsfällen zu führen. Zum anderen beklagt die internationale Forschungs- und Anwendergemeinde das Fehlen effizienter Kommunikationskanäle zwischen den Akteuren im IWRM Umfeld. Letzteres ist allerdings eine grundlegende Voraussetzung, um bestmögliche Lösungsansätze für die bestehenden und zukünftigen Herausforderungen der Wasserbewirtschaftung zu erarbeiten.

Die vorliegende Arbeit verfolgt in dieser Hinsicht einen doppelten Ansatz. Einerseits wird eine IWRM Modellierung- und Szenarien-Planung auf Teileinzugsgebiets-Skala durchgeführt und untersucht. Andererseits wird ein Ansatz entwickelt, der ein kollaboratives und verteiltes Management von planungs- und entscheidungsrelevantem Wissen mittels semantischer Webtechnologie unterstützt. Bei beiden Ansätzen erfolgt die Erarbeitung exemplarisch für das Jordanische Wadi Shueib (~190 km²).

Das Wadi Shueib ist ein dicht bevölkerten Seitental des Jordans, in welchem Nutzungskonflikte durch vorherrschende Versorgungsknappheit sowie inadäquate Abwasseraufbereitung und -ableitung, grenzwertüberschreitende Belastungen vorhandener Ressourcen und bisher nicht umgesetzte Schutzzonenkonzepte gegeben sind. Eine umfassende Analyse der bei lokalen Entscheidungsträgern vorhandenen Datensätze erlaubte die Erstellung holistischer Wasserbilanz-Zeitreihen auf monatlicher Skala. Das erworbene Systemverständnis wurde in einem WEAP Model repräsentiert. Zielstellungen der nationalen Wasserstrategie und geplante Umsetzungsvorhaben dienten als normative Richtlinien zur Erarbeitung einer Reihe von Planungsalternativen und Bewertungsindikatoren für die lokalen Herausforderungen im Wadi Shueib. In beispielhaften Szenarien-Simulationen bis 2025 wurde aufgezeigt, dass die derzeitigen Umsetzungsvorhaben möglicherweise nationale Planungsziele verfehlen. Verstärkte Anstrengungen erscheinen hier insbesondere im Bereich des Ressourcen-Schutzes und bei der Reduzierung der hohen Versorgungsverluste vielversprechend. Der vorgestellte Modellierungs-Ansatz zeigt Möglichkeiten auf, Entscheidungsträger auf der lokalen Planungsebene zu unterstützen und dabei ebenso nationale IWRM-Prinzipien sowie ein wissenschaftlicher Methodik gerecht zu werden.

Die Wadi Shueib Modellierungsstudie diente auch als Vorlage, um das Potential semantischer Technologien für ein IWRM-Wissensmanagement zu untersuchen. Dazu wurde ein Wissensmanagement-Konzept erstellt und eine Bedarfsanalyse durchgeführt, sowie der Prototyp einer Ontologie für IWRM Planungs- und Entscheidungswissen formalisiert. Eine anfängliche Arbeitshypothese bestand in der Annahme, dass eine semantische Strukturen zu semi-automatisierten IWRM-Analyse-Systemen, beispielsweise nach dem bekannten DPSIR-Model (Driver, Pressure, State, Impact, Response), führen könnten. Im Laufe der Arbeit konnten Grenzen und Problematik generischer Wissensrepräsentation mittels klassischer Modelle untersucht und aufgezeigt werden. Das Konzept wurde schließlich mittels eines semantischen Wikis implementiert und der Forschungs- und Anwendergemeinde des SMART-Projektes zur Verfügung gestellt. Die Plattform zeigt deutlich, wie zentrale Wissensmanagement-Anforderungen in einem kollborativen semantischen Wiki unterstützt werden können. Beispielsweise können Wassersektor Experten dabei unterstützt werden ihre Arbeiten und Analysen innerhalb einer semantischen Struktur zu dokumentieren und zu formalisieren, mit Kollegen gemeinsam daran zu arbeiten, konsistent mit anderen Arbeiten zu verknüpfen und ihre Erkenntnisse dauerhaft und sichtbar einem großen Publikum über das Internet zur Verfügung zu stellen. Zukünftige IWRM-Initiativen könnten von vergleichbaren Ansätzen profitieren und die vorgelegte Arbeit bietet als Pilotstudie wertvolle Einsichten zu diesem Thema.

Für meine Mutter Evelyn.

Acknowledgements

This thesis was prepared at the Institute of Applied Geosciences at the Karlsruhe Institute of Technology and in large parts funded in the framework of the SMART-Project by the German Federal Ministry for Education and Research. I am grateful for the opportunity this constellation has granted me. And there are more than a few people whom I have to sincerely thank for their support during my journey towards this thesis.

The first thanks of course are appropriate for those whose work made my work possible. Especially, Prof. Dr. Heinz Hötzl whose long years of experience, commitment and scientific excellence not only unlocked the doors to research in the Middle East, but also provided guidance and ideal. And Dr. Leif Wolf who initiated many of the fundamental ideas that could be refined in this thesis and who not only became an influential teacher, but a dear friend, along the way. And also Prof. Dr. Nico Goldscheider for his continuing support and valuable advice in the late and critical phase of this work.

Large parts of this thesis also relied on the help of various people at the Ministry of Water and Irrigation, the Water Authority and the Jordan Valley Authority in Jordan for the information and data they supplied me with. And here especially Eng. Ali Subah, Eng. Nayef Seder, Dr. Khair Hadidi and Eng. Ahmad Ulimat who helped me through the jungle of Jordanian water sector data. But there have been many more.

And very special thanks to Benedikt Kämpgen from the Institute of Applied Informatics and Formal Description Methods at the KIT, who joined me on the task to find formal structure in water system knowledge and provided important and clever assistance to my work. In this regard I also want to thank Dr. Denny Vrandečić on behalf of the whole SMW community for their brilliant work.

I would like to thank my SMART colleagues from Germany, Israel, Jordan and Palestine for the pleasant and inspiring, formal or informal discussions on our regular meetings. And here especially my fellow PhD-Students (or already PhDs), who formed a beautifully international family of fine and good-humoured souls.

I would like to express very warm thanks to my wonderful colleagues in Karlsruhe, those whose work was directly connected with mine through SMART: Dr. Jochen Klinger, Dr. Wasim Ali, Paulina Alfaro, Felix Grimmeisen, Julian

Xanke, Dr. Julia Sahawneh, Dr. Amani Alfarra, Stefanie Kralisch, but all the other good people at the Institute no less.

I would also like to especially thank my dear friends: Moritz Zemann and Dr. Dino Klotz who so very often offered me shelter, when office hours had been too long again, Anne Schwab, who so very often motivated me with her belief in me. Dr. Burnd Chapligin, Frank Benkert, Jonas Laisney, Miriam & Tobias Wirsing, Christian Groth and Christopher Tracy who helped me so much when I needed it most, and Joanna Dornia who supported me selflessly during the hardship of the last meters. And there have been so many more.

My family and friends are everything and without them I would be nothing. Their love, understanding and support have been the essential ingredients in this thesis. I thank you. Forever.

Table of Contents

Abstract i

Kurzfassung iii

Acknowledgements vii

Table of Contents ix

List of Figures xv

List of Tables xvii

List of Abbreviations xix

1. Introduction 1

    1.1. Challenges of the Global Water Situation 1

    1.2. Motivation of this Thesis 4

    1.3. Objectives 5

    1.4. Thesis Structure 6

2. General Background 9

    2.1. Water Scarcity in the Lower Jordan River Basin 9

        2.1.1. Geographic Overview 9

            2.1.1.1. Geography 9

            2.1.1.2. Climate 10

            2.1.1.3. Ecology and Land Use 10

            2.1.1.4. Geology and Hydrogeology 12

        2.1.2. Today's Water Situation in the Lower Jordan River Basin 13

        2.1.3. History of Conflict and Cooperation in the Jordan River Basin 16

        2.1.4. Institutional Setting in the Jordanian Water Sector 17

    2.2. Integrated Water Resources Management 21

    2.3. Decision Support 27

        2.3.1. IWRM Planning Frameworks 28

2.3.2. Indicator Frameworks 29

2.3.3. Decision Support Systems 31

2.3.4. Scenario Planning 34

2.3.5. Summary 38

2.4. Knowledge Management 40

2.4.1. What is Knowledge Management..40

2.4.2. Defining Knowledge in Knowledge Management..............................41

2.4.3. Knowledge Management Instruments...44

2.4.4. Semantic Knowledge Representation Structures.............................47

2.4.5. Knowledge Management with Wikis and Semantic Wikis 49

3. IWRM-Modelling and Scenario Planning...57

3.1. Framework for IWRM-Modelling and Scenario Planning 57

3.2. Study Area 58

3.2.1. Selection of the Study Area 58

3.2.2. Geographic Overview 59

3.2.3. Climate 60

3.2.4. Geology 63

3.2.4.1. Geological Setting...63

3.2.4.2. Stratigraphy...65

3.2.5. Hydrogeology 67

3.2.5.1. Wells..70

3.2.5.2. Springs 71

3.2.6. Hydrology 74

3.2.6.1. Catchment Drainage and Runoff...74

3.2.6.2. Reservoir 75

3.2.7. Population and Land Use 77

3.2.8. Water Supply and Sanitation 78

3.2.9. Water Sector Challenges in the Wadi Shueib..................................80

3.3. IWRM-Model for the Wadi Shueib...81

3.3.1. Modelling Approach 81

3.3.2. The WEAP21 Water Evaluation and Planning Software...................84

3.3.3. Available Data...86

3.3.4. Data Quality Control...87

3.3.5. Data Pre-processing...90

3.3.5.1. Areal Rainfall...90

3.3.5.2. Areal Evapotranspiration...93

3.3.5.3. Surface Runoff...98

3.3.5.4. Groundwater Recharge...102

3.3.5.5. Water Consumption and Return Flows 105

3.3.6. Wadi Shueib Water Balance...107

3.3.7. Wadi Shueib WEAP-Model...110

3.3.7.1. General Model Characteristics.................................110

3.3.7.2. Schematic WEAP Model...110

3.3.7.3. Model Calibration and Validation.........................113

3.3.7.4. Model Uncertainty...117

3.4. Wadi Shueib Planning Scenarios 119

3.4.1. Objectives and Plans of the Jordanian Water Strategy....119

3.4.2. IWRM Performance Indicators 122

3.4.2.1. I.1a/b: Municipal Waste Water Treatment/Recharge Ratio.......124

3.4.2.2. I.2: Available Renewable Water for Internal Use....................125

3.4.2.3. I.3: Water Supply Shortage Index.................................126

3.4.2.4. I.4: Water Supply Requirement.................................126

3.4.2.5. I.5: Full Water Service Cost 127

3.4.2.6. I.6: Environmental Water Stress (with return flows) 128

3.4.3. Alternative Development Scenarios 129

3.4.4. Driving Forces...131

3.4.4.1. Precipitation...131

3.4.4.2. Temperature and Evapotranspiration 133

3.4.4.3. Population 134

3.4.4.4. Water Demand 136

3.4.4.5. Internal Driving Forces 139

3.4.5. Scenario Simulation Results and Discussion 143

3.4.5.1. I.1a/b: Municipal Waste Water Treatment / Recharge Ratio.....143

3.4.5.2. I.2. Available Renewable Water for Internal Use....................145

3.4.5.3. I.3: Water Supply Shortage Index.................................148

3.4.5.4.   I.4: Water Supply Requirement 151

3.4.5.5.   I.5: Full Water Service Cost 153

3.4.5.6.   I.6: Environmental Water Stress 155

3.4.6.   Summary and Discussion of the Scenario Simulations 156

3.5.   Summary 159

3.6.   Conclusions 160

3.6.1.   Water Accounting for the Wadi Shueib Catchment 160

3.6.2.   The Wadi Shueib WEAP Model as an IWRM Modelling Tool 161

3.6.3.   IWRM Planning in the Wadi Shueib 162

3.6.4.   Results of the Scenario Simulation 163

4.   Knowledge Management for IWRM Plannig and Decision Support 167

4.1.   Introduction 167

4.2.   Perception of KM in the Jordan Valley Water Sector 170

4.3.   Requirements Analysis 172

4.3.1.   Terminology 172

4.3.2.   Knowledge Management Design Goal 173

4.3.3.   Challenges of IWRM Knowledge Representation 175

4.3.3.1.   Collaborative Ontology Engineering 175

4.3.3.2.   Level of Formalization 176

4.3.3.3.   Definition of the IWRM Knowledge Domain 177

4.3.3.4.   Knowledge Representation with Semantic MediaWiki 177

4.3.4.   Use Cases 178

4.3.4.1.   UC-1: Create IWRM-Entity 178

4.3.4.2.   UC-2: Document Expert Analysis 178

4.3.4.3.   UC-3: Document IWRM-Process 178

4.3.4.4.   UC-4: Contribute to Existing IWRM-Entities 179

4.3.4.5.   UC-5: Reuse Knowledge 179

4.3.4.6.   UC-6: Retrieve Analyses for a Study Area and Subject 179

4.3.4.7.   UC-7: Link IWRM-Entity to Representation in Database 179

4.3.5.   Semi-formal IWRM-Ontology 180

4.3.6.   Requirements 181

4.3.7.   Semantic MediaWiki 182

4.4. Implementation 185

4.4.1. Hardware and Software 185

4.4.2. Architecture 186

4.4.2.1. Dropedia-Ontology 186

4.4.2.2. User Interface and Interaction Architecture 188

4.4.2.3. Connection to the External Project Database 190

4.4.3. Use Case Workflows 191

4.5. Evaluation 194

4.5.1. Knowledge Representation Requirements 194

4.5.2. Usability Requirements 195

4.5.3. Functional Requirements 196

4.6. Summary 198

4.7. Conclusions 199

4.7.1. Introducing Semantic Knowledge Management to IWRM ... 199

4.7.2. Expressive IWRM-Knowledge Representation with Ontologies 200

4.7.3. Semantic MediaWiki for Collaborative IWRM Knowledge
Management 200

5. Research Summary and Outlook 205

5.1. Outlook and Recommendations 206

5.2. Critical Remark: Sharing Knowledge – Experiences from DROPEDIA .. 208

6. References 211

7. Appendix 237

List of Figures

Fig. 1.1: Areas of physical and economic water scarcity. ... 2

Fig. 2.1: The Jordan River Basin and the Dead Sea Basin. ... 11

Fig. 2.2: History of the Jordanian water sector. ... 18

Fig. 2.3: Principles-Processes-Tools in the IWRM framework. ... 25

Fig. 2.4: Phases and steps in the IWRM decision process ... 28

Fig. 2.5: Framework of the DPSIR-model. ... 30

Fig. 2.6: Scenario Funnel ... 35

Fig. 2.7: The Data-Information-Knowledge-Wisdom-Pyramid ... 41

Fig. 2.8: Tools in the Knowledge Management landscape ... 46

Fig. 2.9: Difference between non-semantic and semantic wikis ... 52

Fig. 2.10: Building blocks of a semantic wiki knowledge structure. ... 53

Fig. 3.1: Framework of the Wadi Shueib-planning approach ... 57

Fig. 3.2: Location and Overview of the Wadi Shueib study area. ... 59

Fig. 3.3: Long term average annual precipitation in the Wadi Shueib. ... 60

Fig. 3.4: Average temperature and precipitation ... 61

Fig. 3.5: Long term interannual variability of average areal rainfall ... 62

Fig. 3.6: Monthly averages of Class A-Pan evaporation rates ... 62

Fig. 3.7: Geological Map of the Wadi Shueib catchment area ... 64

Fig. 3.8: Geologic profile through the study area ... 65

Fig. 3.9: Registered springs and groundwater wells in the Wadi Shueib ... 70

Fig: 3.11. Annual and monthly discharges of the Wadi Shueib springs ... 72

Fig. 3.10: Small and large spring in the Wadi Shueib area. ... 72

Fig. 3.12: Recorded rainfall at As-Salt station and inflow at the Reservoir ... 75

Fig. 3.13: Wadi Shueib Dam and Reservoir. ... 76

Fig. 3.14: Observed monthly runoff storage in the Wadi Shueib Dam ... 76

Fig. 3.15: Frequency of waste water tank emptying in As-Salt ... 79

Fig. 3.16: Workflow of the IWRM modelling and scenario planning study. ... 83

Fig. 3.17: WEAP21 conceptual water balance elements ... 85

Fig. 3.18: Neighbouring similar rainfall gauges ... 89

Fig. 3.19: Annual areal precipitation in the Wadi Shueib catchment area ... 92

Fig. 3.20: (a) Average ET0 and Pan evaporation and (b)+(c) Scatterplots ... 96

Fig. 3.21: Separation of the Wadi Shueib Catchment ... 97

Fig. 3.22: Areal precipitation and reference evapotranspiration time series ... 97

Fig. 3.23: Base flow separation with the WHAT Eckhardt Filter module ... 100

Fig. 3.24: Groundwater Contribution Zones of the four major springs ... 103

Fig. 3.25: Water Balance for Wadi Shueib ... 109

Fig. 3.26: Schematic representation of the Wadi Shueib WEAP-model 111

Fig. 3.27: Calibration and validation results 116

Fig. 3.28: Schematic of the topics addressed in Jordan's Water strategy 120

Fig. 3.29: Alternative development scenarios for the case study 131

Fig. 3.30: Time series of interpolated annual areal rainfall volumes 132

Fig. 3.31: Projected precipitation time series 133

Fig. 3.32: Population scenario projections 135

Fig. 3.33: Per capita municipal demand development 137

Fig. 3.34: Industrial demand development 138

Fig. 3.35: Waste water treatment ratio and volume of untreated waste water .. 145

Fig. 3.36: Resources for groundwater, stored surface runoff, reclaimed water .. 148

Fig. 3.37: Municipal water shortage index and water imports needs. 150

Fig. 3.38: Development of the agricultural water supply shortage index 151

Fig. 3.39: Municipal water supply requirement 152

Fig. 3.40: Full service cost of municipal water 154

Fig. 3.41: Environmental water stress 155

Fig. 3.42: Summary of the standardized scenario simulation results 158

Fig. 4.1: Components of the SMART-Decision Support framework 169

Fig. 4.2: Expert opinion poll about transparency, participation and
knowledge management in the Lower Jordan River Valley 171

Fig. 4.3: Concepts and user roles in the IWRM knowledge process. 174

Fig. 4.4: Class-structure of the semi-formal IWRM-ontology. 180

Fig. 4.5: DROPEDIA welcome screen. 185

Fig. 4.6: Subset of the DROPEDIA-Ontology 188

Fig. 4.7: User interface architecture of DROPEDIA 189

Fig. 4.8: Subset of the form and template structure of DROPEDIA 190

Fig. 4.9: Architecture of DROPEDIA/SMART-DB connection 191

Fig. 4.10: DROPEDIA workflow for IWRM-Object and Analysis contribution. 193

List of Tables

Table 2.1: Available freshwater resources in the Lower Jordan River Basin 14
Table 2.2: Selection of IWRM definitions 23
Table 2.3: Comparison of the sequential scenario development phases 38
Table 2.4: Strengths and limitations of wikis 51
Table 3.1: Stratigraphy of the Wadi Shueib 66
Table 3.2: Hydrogeological classification in Wadi Shueib 69
Table 3.3 List of groundwater-wells in Wadi Shueib 71
Table 3.4: Demographic Development in Wadi Shueib between 1994 and 2004 .. 77
Table 3.5: Operation data from the Wadi Shueib waste water treatment plants. 79
Table 3.6: Overview of the available data 87
Table 3.7: Comparison of monthly and annual areal rainfall interpolations 93
Table 3.8: Comparison of the interpolation results and literature values 93
Table 3.9: Optimized Filter Parameter and BFI_{max} for base flow separation 100
Table 3.10: Estimated monthly rainfall-runoff for three water years. 101
Table 3.11: Recharge balance for the Wadi Shueib springs 105
Table 3.12: Return flow balance for As-Salt and Fuheis/Mahis 107
Table 3.13: Model performance indicators 116
Table 3.14: Objectives and approaches of the Jordanian Water Strategy 121
Table 3.15: Jordanian population growth rate projections 135
Table 3.16: Development of Jordan's sectoral demand and available resources. 136
Table 3.17: Development of the model parameters 140
Table 3.18: Investment assumptions for the Wadi Shueib planning scenarios ... 143
Table 3.19: Development of the municipal waste water treatment ratio 144
Table 3.20: Volume of untreated waste water discharge and ratio to
                groundwater recharge 144
Table 3.21: Per capita volume of internal groundwater resources 146
Table 3.22: Per capita volume of internal surface water resources 147
Table 3.23: Per capita volume of internal reclaimed water 147
Table 3.24: Unmet municipal demand and water supply shortage index 149
Table 3.25: Unmet agricultural demand and water supply shortage index 150
Table 3.26: Water supply required for meeting the anticipated demand 152
Table 3.27: Full water service costs for the municipal sector 153
Table 3.28: Evaluation matrix of standardized indicator scores 157
Table 4.1: Terminology of the representational units used for knowledge
                modelling 172

Table 4.2: Requirements for a collaborative knowledge management platform for IWRM planning and decision support. 181

List of Abbreviations

AHP	Analytic Hierarchy Process
BAU	Business-as-Usual-Scenario
DIKW	Data-Information-Knowledge-Wisdom hierarchy
DPSIR	Driver-Pressure-State-Impact-Response
DSS	Decision Support System
DWWT	Decentralized Waste Water Treatment
FI	Full-Implementation-Scenario
GWCZ	Groundwater Contribution Zone
HRP	High Resources Pressure-Scenario
IWRM	Integrated Water Resource Management
JD	Jordanian Dinar
KB	Knowledge Base
KM	Knowledge Management
LJRB	Lower Jordan River Basin
LJV	Lower Jordan Valley
LRP	Low Resource Pressure-Scenario
MCM	million cubic meter
OWL	Web Ontology Language
RDF (S)	Resource Description Format (Schema)
SMART	Sustainable Management of Available Resources with Innovative Technologies
SMW	Semantic MediaWiki
SPARQL	SPARQL Protocol And RDF Query Language
TWW	Treated Waste Water
WEAP	Water Evaluation And Planning System
WWTP	Waste Water Treatment Plant
XML	Extensible Markup Language

The White Rabbit put on his spectacles.
"Where shall I begin, please your Majesty?" he asked.
"Begin at the beginning,", the King said, very gravely,
"and go on till you come to the end: then stop."
Lewis Carroll (1865: Alice's Adventures in Wonderland)

1. Introduction

1.1. Challenges of the Global Water Situation

Fundamental change and rapid development of human society during the last two centuries has put immense pressure and competition, be it in quantitative or qualitative regard, on the natural freshwater resources in numerous regions of the world. Future water security is one of the major challenges addressed by the Millennium Development Goals, not only in terms of safe drinking water supply and sanitation, but also in its inherent relation to food and health security, economic development and environmental conservation. Typically, countries affected by the most obvious and dramatic negative effects are among the world's poorest. Their water stress usually caused by limited natural freshwater availability or by the lack of adequate infrastructure to utilize available resources - or most often even both.

The United Nations define "water stress" to occur where the volume of freshwater available for supply to agricultural, industrial, environmental and domestic purposes lies below 1700 m³ per year and person, and "water scarcity" occurs when annual water supplies drop below 1000 m³ per person (Falkenmark, 1989). Many regions in the World experience even much more critical scarcity, providing less than 500 m³ per person and year, which is then considered as "absolute water scarcity". Global studies estimate that today approximately one billion people are already experiencing water stress or water scarcity due to deficient resources (UNDP, 2006). Additionally, 500 million more live in regions threatened by water stress, due to their natural surface water resources being exploited beyond sustainable limits (CAWMA, 2007). As illustrated in the map (**Fig. 1.1**), physical scarcity appears to be most acute in countries of the Middle East and Northern Africa, but also in parts of Northern China, India, Australia and the United States. Another 1.6 billion people face shortage of clean freshwater even in countries with abundant resources, due to the lack of appropriate infrastructures and insufficient capacities to develop such (CAWMA, 2007). Nearly a quarter of Sub-Saharan Africa's population is estimated to live under such conditions of economic water scarcity (UNDP, 2006).

Beyond natural hydrologic conditions, lacking infrastructure and high population pressure, another major reason for the degradation of freshwater resources often

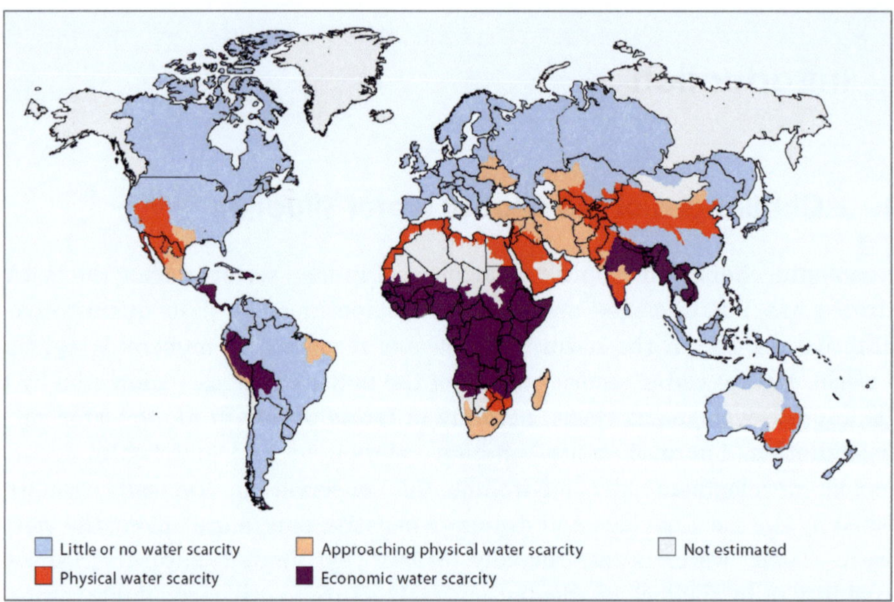

Fig. 1.1: Areas of physical and economic water scarcity. Physical water scarcity as defined here occurs when 75% of the total available water resources are withdrawn for utilization. The threshold for approaching physical scarcity is 60%. Economic water scarcity occurs in regions where malnutrition exists despite abundant resources. (UNEP/GRID-Arendal, 2008).

lies in mismanagement. Notwithstanding the rising global awareness for sustainability, the rapid development of urbanisation, industrialisation and economic growth has ever outpaced society's adaptive reactions to the complex demands of future water security. Thus, inefficiently targeted investment, insufficient human capacity, ineffective institutions and poor governance have resulted in continuous misuse of the resources over the past decades. Extensive policy and market failures have received only limited corrections from concerned institutions (Biswas et al., 2009).

Unfortunately the countries that are facing the most urgent need for institutional reform and large investments are again the world's poorest in terms of per capita income. Grey and Sadoff (2007) see these countries in a low-level equilibrium trap, having never been able to make the investments needed to achieve water security, because such investments can only be resourced from the growth that water insecurity itself constrains. From a slightly more pessimistic perspective, this situation does not only pose an equilibrium trap but an actual downward spiral, since the driving factors that are denying improvements are also likely to induce negative feedback on the actual water situation. This is the case for example with the process of desertification or the salinization of groundwater due to continuous overpumping. Once more, the consequences are persistent in limited development opportunities, environmental degradation, declining groundwater tables and increasing conflict over water allocations.

The highest social and political complexity in water management matters arises where competition over water resources becomes transboundary. As today almost half of the world's population lives in internationally shared river basins, bi- or multilateral collaboration over water resources truly is an issue of global significance. History has revealed that interaction on transboundary freshwater resources bears as much, or actually even more, incentives for cooperation and political stability, as it holds potential for conflict if neglected (Hamner & Wolf, 1998). However, many challenges remain to be addressed in order to extend and stabilize sustainable water sharing between riparian actors, particularly with regard to an expected future of worldwide intensifying freshwater competition.

Despite the international efforts, the envisioned Millennium Development Goals for water security appear unlikely to be met in 2015 (UN, 2010a). On the contrary, several trends are likely to aggravate the situation over the coming years in fast pace. Many of the already water-stressed countries are facing continuous high population growth rates. In addition, the per capita water use has been growing ever faster, while water use patterns changed. Simultaneously, pollution and decline of natural freshwater bodies show upward trends.

These considerations lead to the prognosis that the pressure on water resources will constantly increase with rising demands from agricultural, municipal, industrial and environmental usages. By 2025 it is estimated that 1.8 billion people will be living in regions with severe water scarcity, and two-thirds of the world population could be under conditions of water stress (Alcamo et al., 2003; Oki & Kanae, 2006; UN-Water & FAO, 2007). Climate change scenarios expect changes in rainfall patterns and amplification of extreme events and additionally undermine the urgent need for a more coordinated and integrated approach to the present water resources situation (IPCC, 2007; Oki & Kanae, 2006).

In the effort to cope with these challenges, the Integrated Water Resources Management approach has been the most prominent strategic water management concept in the international discourse during the last decade. But its implementation appears an intricate challenge in itself. In this respect, a uniform tenor across disciplines is heard, which consistently emphasises the necessity to enhance the communication and understanding between scientific domains, as well as between scientists, policy makers and the public society, in order to ensure a best-possible response to the upcoming water resources challenges (Jacobs et al., 2005; Liu et al., 2008; Oki et al., 2006; Pahl-Wostl, 2007).

1.2. Motivation of this Thesis

Following the paradigm of Integrated Water Resources Management (IWRM) presumes a basin-wide holistic modelling approach and the definition of sustainable long-term planning scenarios. Therefore, a comprehensive IWRM framework has to combine complex data and information from various different domains to produce transparent and application-ready information for decision making. These requirements result in two fundamental challenges, which are currently not yet practically resolved:

Basin scale planning models and scenarios are often experienced as useful for setting broad national water management strategy goals, but they frequently lack reliability due to the non-validated assumptions on their local implementation. This certainly contributes to the main critiques of the IWRM approach as a concept that is rather ideological and vague (Biswas, 2004; Lankford et al., 2007) and generally without sufficient empirical evidence for its benefits (Galaz, 2007; Jeffrey & Gearey, 2006; Merrey, 2008). In this regard, it has been suggested that the theoretical construct of IWRM is not to be understood as universal operational guideline, but rather at a normative and strategic level (Mitchell, 2004) and that any local or regional initiative eventually has to develop its own specific practice adapted to the given situation (GWP, 2000). Methodical approaches that support planners and decision makers in finding appropriate balance between the local planning reality, national IWRM policies, and a sound scientific basis, are still subject of research in international IWRM initiatives.

The other discussion, that has already been progressing for several years, concerns the gap between environmental scientific modelling and respective political decision making (Acreman, 2005; Cash et al., 2002; Kolkman et al., 2005; Liu et al., 2008). Most contributions on that topic agree on the urgent need of new and improved communication channels between science and policy. Regarding multi-disciplinary research projects, it also the information flow between the participating experts that poses challenges. The combination of physical system modelling with social, ecological and economic factors makes an IWRM analysis a highly collaborative and knowledge intensive process. And the impacts of planning scenarios have to be jointly assessed from various perspectives, corresponding to the knowledge of experts in different domains as well as to the interests of various stakeholders. Thus, the development of a comprehensive and interdisciplinary water resources knowledge base is perceived as fundamental to integrated water resources assessment and consequent decision making (GWP, 2000). In this regard, it is also recognised that contemporary knowledge management techniques can contribute to improving the performance and effectiveness of both capacity development (Luijendijk & Lincklaen Arriëns, 2007) and knowledge sharing (Giupponi & Sgobbi, 2008; Roux et al., 2006) in the water sector, provided that there is a basic capacity in place to coordinate this approach. However, the direct support of IWRM decision processes in the planning practice through contemporary knowledge management strategies is still lacking satisfactory consideration in current IWRM planning and decision support.

1.3. Objectives

This thesis mostly evolved from the work within the SMART-Project for Sustainable Management of Available Water Resources with Innovative Technologies in the Lower Jordan Valley (Wolf & Hötzl, 2011). Against this background, it is as much an input to a multi-lateral and interdisciplinary IWRM research initiative, as a contribution to the domain of IWRM planning and decision support.

Within this framework, the primary objectives were to

1. Establish and test an operational approach for sub-basin-scale IWRM modelling and scenario planning within a case study for the Jordanian Wadi Shueib sub-catchment.
2. Explore the impacts of the current water resources planning practice in the Wadi Shueib and compare them to the national strategic objectives.
3. Investigate the possibilities to make use of contemporary semantic knowledge management techniques to improve collaboration in operational IWRM planning and decision making.
4. Develop a functional platform that can be used for collaborative knowledge management by the participants and stakeholders of the SMART project, and has a potential to be integrated into future IWRM initiatives.

These objectives brought forth a set of essential research questions that are not yet answered in the international research literature:

- Is the currently available data basis for the Wadi Shueib catchment sufficient to establish an operational sub-catchment IWRM-model?
- How can planning alternatives be practically evaluated and compared in terms of IWRM principles and the Jordanian national policies?
- What are the critical knowledge management requirements in IWRM planning and decision support in Jordan?
- How can knowledge of participants in the IWRM-process be practically formalized to allow collaborative authoring with semantic technologies?
- How can a collaborative knowledge management approach integrate into the workflow of participants in the Lower Jordan Valley IWRM-process?

The presented thesis delineates the process of the conceptual development and application of the Wadi Shueib WEAP IWRM-modelling and scenario planning study, as well as the parallel design and implementation of the DROPEDIA semantic wiki platform for collaborative IWRM knowledge management.

1.4. Thesis Structure

Chapter 2 begins with an introductory overview to the general geography of the Lower Jordan River Basin, and then establishes a context of the region's water resources related challenges in a short review of the historical and contemporary situation. With a comprehensive literature review, the following sections embark on a thorough attempt to elaborate a clearer understanding of the broadly defined concepts of **2.2** Integrated Water Resources Management, **2.3** Decision Support and Scenario Planning and **2.4** Knowledge Management.

Chapter 3 is dedicated to the development and application of the IWRM-modelling and scenario planning framework and the Wadi Shueib use case. It describes the water resources related aspects of the **3.2** Wadi Shueib study area, the **3.3** development of the Wadi Shueib WEAP model and the **3.4** Planning Scenario development, simulation and result discussion. A **3.5** summary and the drawn **3.6** conclusions are given at the end.

Chapter 4 introduces the approach of collaborative knowledge management for IWRM planning and decision support by shortly discussing the **4.2** perception of knowledge management in the Lower Jordan Valley water sector. The development of the knowledge management platform includes a preliminary **4.3** requirements analysis to guide the subsequent **4.4** implementation of the DROPEDIA semantic wiki platform, and a concluding **4.5** evaluation against the requirements. Again, **4.6** summary and **4.7** conclusions are given at the end.

Chapter 5 gives a brief recollection of the most important aspects found in this thesis and concludes with an outlook on potentially interesting future studies.

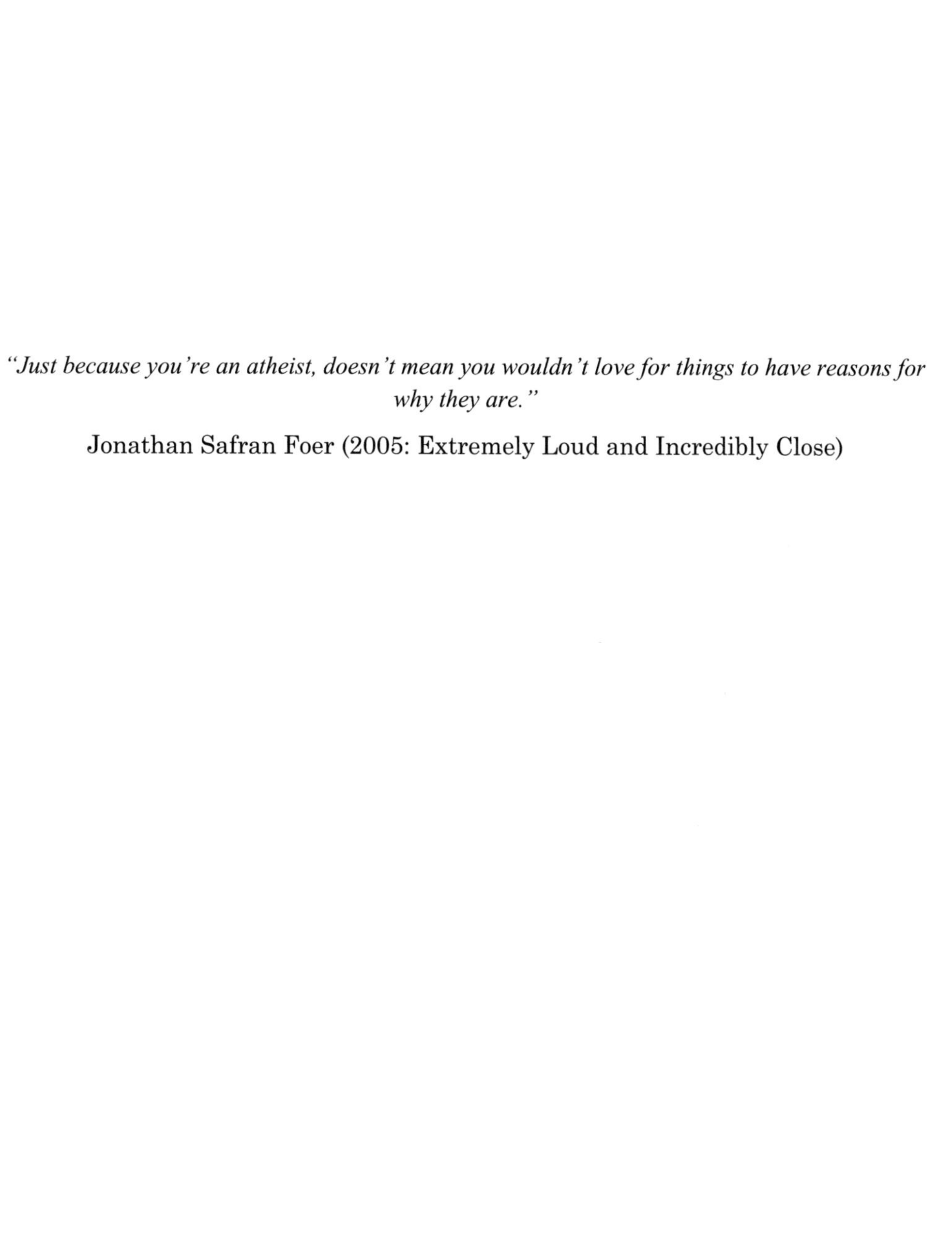

"Just because you're an atheist, doesn't mean you wouldn't love for things to have reasons for why they are."

Jonathan Safran Foer (2005: Extremely Loud and Incredibly Close)

2. General Background

2.1. Water Scarcity in the Lower Jordan River Basin

Countries in the Middle East are among those suffering the highest grade of water scarcity worldwide (CAWMA, 2007). The situation appears particularly critical along the Jordan River where utmost physical water scarcity meets very high population growth and development rates in a region with a constantly sizzling potential for violent conflict.

The following sections will give a general overview on the regional situation in the Lower Jordan River Basin in terms of the natural availability of water resources, the recent history of their utilization and their current status.

2.1.1. Geographic Overview

2.1.1.1. Geography

The catchment of the Jordan River usually is divided into an upper and a lower part. The Upper Jordan River catchment comprises the drainage area of the Jordan River headwaters with its main tributaries (Hasbani, Banias and Dan) originating in the Mount Hermon region of the south-western Anti-Lebanon mountain range and reaches until the Jordan River enters into the Sea of Galilee (*also: Lake Tiberias*), the latter usually also included in the upper catchment part. The upper catchment is shared between the southern parts of Syria and Lebanon and the northern part of Israel. The **Lower Jordan River Basin** (LJRB) as a part of the Jordan Rift Valley comprises the Lower Jordan River Valley (LJV) between the Sea of Galilee and the Dead Sea as well as the eastern and western river valley escarpments and the Yarmouk and Zarqa rivers as main tributaries (**Fig. 2.1**). The whole catchment of the Jordan River from the springs in the Mount Hermon region to the inflow into the Dead Sea has an area of approximately 18,250 km², whereof the LJRB drains about 15,400 km² (MWI, 2004). The Lower Jordan River itself is shared by Israel, Jordan and the Palestinian Westbank as territorial border and mutual water resource.

The Jordan Rift Valley forms an endoreic basin with the Dead Sea as a "dead-end" lake without outlet streams. The rift structure constitutes a distinct topography with terrain elevations below sea level at the Jordan Valley floor (200 m bmsl south of Sea of Galilee and 420 m bmsl at the Dead Sea) climbing in steep ascent to above 1,200 m amsl at the eastern rift flanks (and 900 m on the western side). The entire closed watershed of the Dead Sea covers 40,650 km² (EXACT, 1998), adding the eastern and western wadis that discharge directly into the Dead Sea and the area of the North Wadi Araba Basin to the south. The topographic low of the Jordan Rift valley acts as drainage for the surface water as well as the groundwater from the eastern and western rift flanks.

2.1.1.2. Climate

The mostly arid to semi-arid climate of the LJRB is characterized by winter rainfalls that are mostly related to the Mediterranean circulation. The summer months are usually without any precipitation. The total annual rainfall is increasing from less than 100 mm over the Dead Sea to the higher altitudes of the rift valley escarpments, where it exceeds 600 mm per year in the Mediterranean climate of the Ajloun Mountains. Potential evaporation follows a hereto inverted trend with extreme values of 2,600 mm per year in the Dead Sea area and a maximum of 1,900 mm in the highlands. Actual evapotranspiration losses account for 65 % up to over 90 % of annual rainfall volumes (Flexer et al., 2009; M. J. Haddadin, 2006). The region also experiences strong interannual precipitation variability with a standard deviation in annual totals of about 30 % (Black et al., 2011), that amplifies the uncertainty of available water resources in the region.

2.1.1.3. Ecology and Land Use

The ecosystem of the Lower Jordan River Basin shows a passage from a Mediterranean to semi-desert environment as descending from the rift flanks towards the Jordan River Valley. On the higher altitudes the natural biome includes oak and pine forests, wood- and shrubland which gradually transforms into the dominant xeric shrubland in the Jordan Valley and Dead Sea region. Dense habitats are usually found in the wetlands around natural springs and along the surface water courses with a very high biological diversity and many endemic species (Scott, 1995). Today the Jordan River Basin is characterized by intensive agricultural activity, concentrated as irrigated agriculture along the Jordan Valley Floor and a more rain-fed based on the highlands of the rift shoulders. Some natural forests and reforested areas still cover the upper reaches together with grasslands and orchards of olive and citrus trees that are also quite abundant along the wadis of the rift flanks.

The urban centers are generally situated along the mountain crests of the rift shoulders, whereas smaller communities are found throughout the wadis and the Jordan Valley. During the last decades a thriving tourism sector has evolved and a number of large hotels have been constructed along the Dead Sea shores.

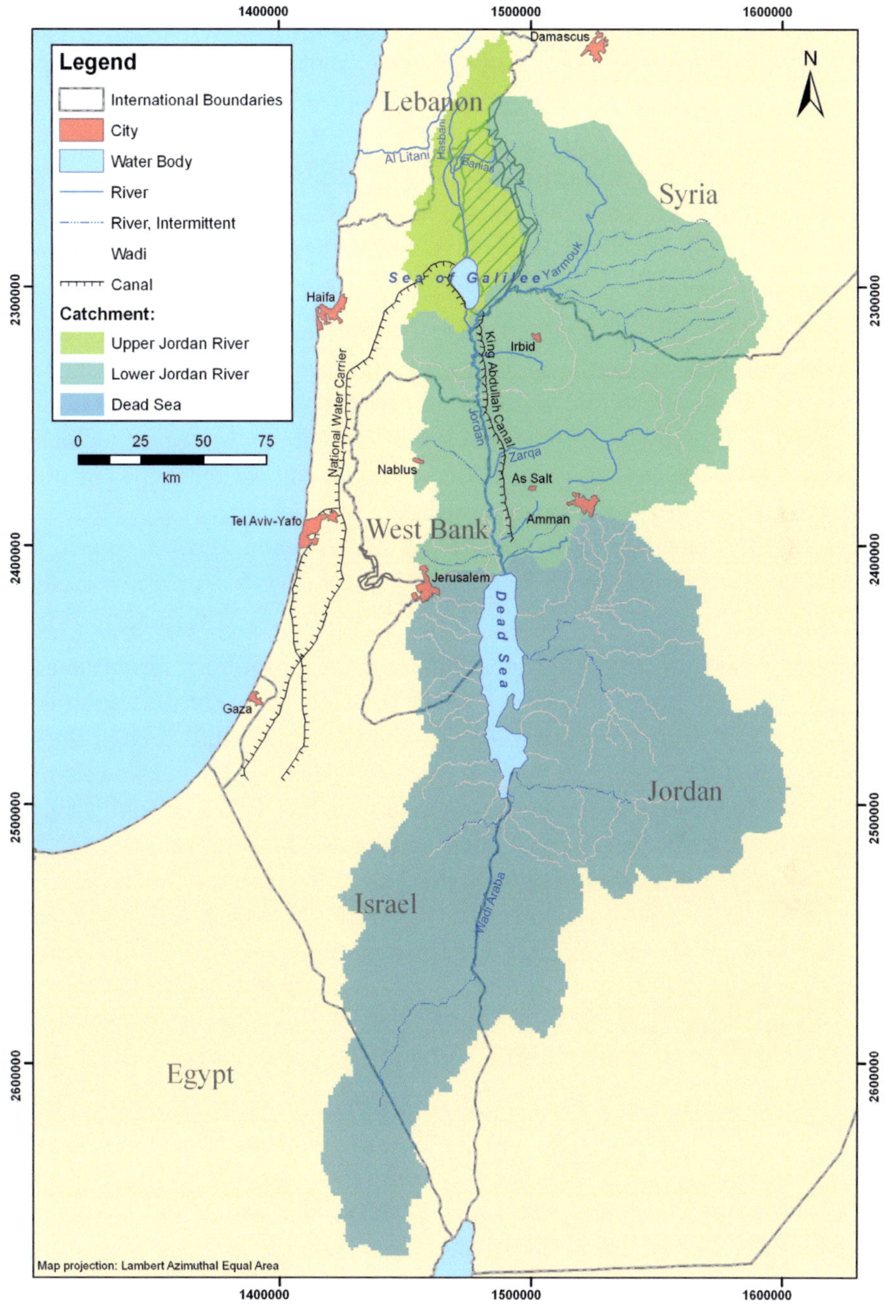

Fig. 2.1: The Jordan River Basin and the Dead Sea Basin.

2.1.1.4. Geology and Hydrogeology

The Jordan Rift Valley is part of the Dead-Sea-Wadi Araba-Transform Valley, a continental transform plate boundary which was formed by the second phase of the break-up of the Arabian-African continent in the Late Cenozoic (Bayer et al., 1988). The Rift Valley formed as a sinistral strike-slip fault system with an estimated 105 km offset connecting the divergent plate boundary in the Red Sea to the convergent plate boundary in the Taurus Mountains in southern Turkey (Freund et al., 1970). For a comprehensive disquisition of the geologic and tectonic history of the Jordan Rift Valley see Horowitz & Flexer (2001).

The lithostratigraphy exposed in the Lower Jordan Rift Valley contains sandstones, carbonates, chert, chalk, evaporates, gravels and sands of Triassic to Holocene age. The eastern and western banks show an asymmetry of successions. The eastern side exposes deeper levels of Triassic to Neogene sediments of limestone, sandstones and marl layers, whereas on the western side the outcrops only start in Late Jurassic. The Jordan Valley is filled with young formations of Miocene to Holocene age that consist mainly of sand, gravel, marls, clays, evaporites and clastic sediments (Bender, 1974).

In the highland areas the limestone and sandstone layers of the Cambrian (southern Jordan) to Upper Cretaceous (northern Jordan) formations build important regional aquifer systems which receive their natural groundwater recharge mainly from precipitation in their outcropping areas along the escarpments in Jordan and the West Bank. The Judea Group on the western side forms the aquifer system of the Mountain Aquifer which is traditionally separated into three basins: the Western Aquifer, the Eastern Aquifer and the North-Eastern Aquifer basins and of which the two latter drain towards the Lower Jordan River Valley. On the eastern side of the Rift the Ram, Zarqa and Kurnub form the Deep Sandstone Aquifer Complex and the Ajloun Group forms the important Upper Cretaceous Aquifer of the Side Wadis Basin. A number of springs emerge from these layers on both sides of the valley which traditionally are an important water source for the communities in the wadis.

In the Jordan Valley itself local aquifers are formed by sand beds and gravel layers in the valley floor deposits and in the alluvial deposits of the side valleys. Recharge to these local aquifers comes from various sources including precipitation, infiltration of wadi drainage, irrigation return flows and upward-seeping water from deep aquifers (Guttman, 2009). Along the foothills of the Jordan Valley several deep salt-bodies with ascending brines as well as some shallow salt bodies result in brackish spring discharges.

2.1.2. Today's Water Situation in the Lower Jordan River Basin

The naturally available water resources in the Lower Jordan Valley consist mainly of waters from Lake Tiberias and the Yarmouk river catchments, smaller streams from the side valley wadis with their base and flood flow and the groundwater bodies of the Mountain Aquifer on the western side and the Jordanian groundwater basins on the eastern side of the Jordan River. Transboundary groundwater and surface water inflows from Syria and Lebanon, but also in between the riparians contribute to their accounted renewable resources. In addition to these conventional resources, the riparians are increasingly utilizing unconventional supplies, e.g. treated wastewater and desalinated water in order to meet their rising demands.

The Lower Jordan River Basin is a naturally water scarce region. Nonetheless, it has experienced an enormous population increase of 654 % in Israel, Jordan and the Palestinian West Bank during the last 60 years from 2.5 million in 1950 to currently more than 16 million people (UN, 2010b). Along with the demographic growth the region has been developing on all sectors, especially in regard to an intensifying agriculture but also growing industries and urban development with rising living standards and a blooming touristic segment. This has resulted in an immense resource-pressure and water has become the most critical natural limitation on the economic growth.

Already today, surface and groundwater freshwater resources are exploited to nearly full extent and in many places the available groundwater is pumped beyond sustainable yield. This has resulted in a continuous degradation of quality and quantity of the surface- and groundwater resources during the last several decades and remaining quantities are constantly depleting and are highly threatened by salinization and contamination (El-Naqa & Al-Shayeb, 2009; Marie & Vengosh, 2001). The most prominent observable effect of the regions environmental water stress is the decline of the Dead Sea surface water table at an average rate of 0.6-1 m/year since the late 1970s (Khlaifat et al., 2010). Salameh & Wl-Naser (1999) estimated the natural water inflow into the Dead Sea prior to the water resources development at 1,980 million cubic meter per year (MCM/a) and the present inflow volume to 617 MCM/a (both figures including fresh and brackish groundwater and surface water inflows from the Lower Jordan River Basin and the Dead Sea Basin (cf. **Fig. 2.1**)).

The majority of runoff in the Jordan River is generated in the northern catchment area of Lake Tiberias and Yarmouk (approx. 1,000 MCM/a) and is for the most part already abstracted by Israel, Jordan and Syria before it reaches the Lower Jordan River. **Table 2.1** gives an approximate water budget for the Lower Jordan River Basin in an average water year.

Table 2.1: Available freshwater resources in the Lower Jordan River Basin (LJRB), current abstractions and additional national resources. The given groundwater volumes comprise renewable and non-renewable resources. Unconventional resources comprise desalination capacities and utilization of treated waste water. Budget estimates given here were compiled from various sources: (MWI, 2004; Nazer et al., 2008; Salameh & El-Naser, 1999; Suleiman, 2004; Weinberger et al., 2012; Zeitoun et al., 2009).

Source	Natural freshwater resources	Abstractions Jordan	Israel	Palestine (West Bank)
		MCM/year		
Upper Jordan and Lake Tiberias	565	55	500	0
Yarmouk inflow from Syria*	400	150	20	0
Eastern LJRB** — surface runoff (flood flow)	100	60	0	0
Eastern LJRB** — groundwater recharge & base flow	200	240	0	0
Western LJRB*** — direct runoff (flood flow)	25	0	5	10
Western LJRB*** — groundwater recharge & base flow	275	0	150	110
Total LJRB freshwater resources	**1565**	**505**	**675**	**120**
Additional resources outside of LJRB — surface		160	150	0
Additional resources outside of LJRB — groundwater		230	800	80
Additional resources outside of LJRB — unconventional		100	675	8
Total national freshwater resources		**995**	**2300**	**208**
Population (estimates for 2010) — *capita*		6187000	7418000	2568555
Per capita available freshwater resources — *m³/c/a*		**161**	**310**	**81**

* Syrian abstraction from the Yarmouk is estimated at currently 200-240 MCM/year (P. M. J. Haddadin, 2006)

** including Jordan Valley and Side Wadis, Yarmouk (Jordanian catchment area), Zarqa River

*** including Jordan Valley and the Eastern & North-Eastern Basins of the Mountain Aquifer

In Jordan for example the current annual average amount of renewable groundwater is estimated at around 275 MCM, but the actual abstraction rate reached 390 MCM in 2010 leading to a severe over-pumping of the aquifers. In addition fossil groundwater resources in the Jafr and Disi basins are currently utilized with a rate of 80 MCM per year (MWI, 2010) and pumping is planned to be increased through the Disi-Amman conveyance project scheduled to operate in 2013. With the contribution of about 530 MCM of exploitable surface water resources as well as 90 MCM of reclaimed water, the overall average available freshwater resources in Jordan are estimated at around 900-1,000 MCM, which is about 160 m³ per capita and day, thus lying considerably below the UN threshold for absolute water scarcity of 500 m³ per person and year and places it as one of the most water scarce countries in the world on a per capita basis.

In terms of water scarcity, the situation in Israel and the Palestinian West Bank is comparable to that in Jordan. In fact, because large parts of the West Bank are regions of higher precipitation and recharge (685 MCM/a estimated average recharge of the Mountain Aquifer), this area could be less stressed in terms of total renewable resources per capita. However, the issue between Israelis and Palestinians regarding the shared resources and their control in the area of the Mountain Aquifer makes it utterly more complicated. With the full utilization of the Tiberias waters and intensive pumping from the Mountain Aquifer and other regional groundwater basins, Israel reaches a per capita availability of 220 m³/c/a of natural freshwater resources which is raised to about 310 m³ per capita through the considerable efforts that are put into the development of unconventional resources (desalination and reclaimed water). In the current situation the Palestinians in the West Bank dwell on an estimated volume of 75-85 m³ per capita of freshwater resources (World Bank, 2009b). When one considers regional differences within the countries and the actual available share of daily drinking water, the situation in the West Bank territories seems even worse (Hareuveni & Stein, 2011; Lein et al., 2001).

In light of the recent trends of population growth and water demand development the current water deficits suffered by the riparians are expected to continuously aggravate. Furthermore, the potential to develop new natural resources is becoming more and more meagre and every additional drop abstracted from the remaining streams flows and aquifer storages means additional stress on the plagued environment of the Lower Jordan River Valley. Thus, the extended exploitation of unconventional resources has become one major development goal for the water sector in the three countries. This, however, has to be seen with regard to the very different background situations, which for example makes large scale desalination projects more difficult for Jordan (due to long-distance and high-altitude conveyance) and almost impossible for the inhabitants of the West Bank. Many hopes, especially in Jordan, are therefore set on the controversial and uncertain realization of mega-projects like the Mediterranean-Dead-Sea or the Red-Sea-Dead-Sea canal (Murakami, 1998). On the other hand, during the last decades the water sector development in the region has already been dominated by a supply-side management (*"greening deserts"*) that focused the majority of efforts on the maximization of available resources. Yet, recently and despite a strong opposition mainly from the agricultural lobby which still is the main water user in the region (agricultural consumption of available freshwater resources: Israel: ~55%, Jordan: ~70%, West Bank: ~70%) it appears that the political will is rising towards urgently necessary demand management approaches (Zeitoun et al., 2012; Zeitoun et al., 2009).

2.1.3.   History of Conflict and Cooperation in the Jordan River Basin

After the declaration of the State of Israel in 1948 the governments in the region began with the pursuit of national plans for the water resources exploitation of the Jordan River Basin (Haddadin, 2002). These unilateral activities soon led to first water related disputes and conflicts between the riparians until, in 1953 and on initiative of the USA, the four countries Lebanon, Syria, Jordan[1] and Israel started working on the so-called "Johnston Plan", with the goal of a multilateral agreement on water use rights for the Jordan River (Smith, 1966). Although the plan was technically accepted, it was never ratified by the Arab League due to politically motivated reasons related to the Arab-Israeli tensions. With the failure of these negotiations, the riparians resumed their unilateral actions in order to secure their share of the Jordan River. In particular Israel, Jordan and Syria continued to develop their water resources and realized major projects in this period that still determine todays water regime in the Lower Jordan catchment (Haddadin, 2002).

During the first half of the 1960s Israel built the National Water Carrier that pumps water from the north-western part of Lake Tiberias to Tel Aviv and into the Negev for irrigation and domestic purposes. At about the same time Jordan realized the East Ghor Canal (today known as: *KAC – King Abdullah Canal*) which diverts water from the Yarmouk River and runs it southward parallel to the Jordan River where the water is used mainly for irrigation in the Jordan Valley and drinking water supply of Greater Amman. In the 1970s Jordan constructed the King Talal Dam and the Wadi al Arab Dam to supplement the KAC system (Haddadin, 2002). Syria constructed several dams on the Yarmouk River and later on also started to utilize spring water from the upper parts of the catchment (Al-Kloub & Abu-Taleb, 1998).

From their very beginning these activities resulted in various smaller military conflicts and contributed to the overall tensions that ultimately reached a climax with the Arab-Israeli War (also: *Six-Day War*) in 1967 and which fundamentally changed the settings in the region through the Israeli occupation of the West Bank and the Golan Heights[2] (Ziegelmayer, 2008). Dispute on water resources questions has ever since continued to put strains on regional politics in the

---

[1] The area of the West Bank was annexed part of Jordan at the time of the Johnston negotiations.

[2] Jägerskog (2003) points out that by these occupations "…Israel enhanced its 'hydro-strategic' as well as its military–strategic position. By the acquisition of two out of the three sources of the upper Jordan River and through increased control of the Yarmuk River,…" and "…including the control over the West Bank aquifers, Israel now had control over 80 per cent of the surface and groundwater resources it uses."

Jordan River Basin and the most recent eruption in form of the *Wazzani conflict* between Israel and Lebanon in 2002 is probably only the latest but not the last outbreak of water related skirmishes in the region (Amery, 2002).

However, despite the persistent potential for conflict, the countries of the Jordan River Basin have also realized their urgent need for cooperation during the last decades. So, after a period of coexisting water resources development without mutual cooperation, the riparians took up serious negotiations during the 1990s and eventually signed several, foremost bilateral, legal agreements on the distribution of the water resources (surface water and groundwater) in the Jordan River Basin (Haddadin, 2002; Jägerskog, 2003). The two central agreements in this respect are the *Israel–Jordan Treaty of Peace* signed in 1994 and the *Israeli-Palestinian Interim Agreement* (also: *Oslo II*) from 1995 (Interim Agreement, 1995; Treaty of Peace, 1994). Although both agreements have been of utter importance to the water question in the region, it has to be stressed that so far they have resulted in different levels of success. The Israel-Jordan Peace Treaty, for all unsolved question and open dispute, has already improved the water situation (especially for Jordan), and even fostered collaboration beyond the original agreements between the two countries. Whereas the negotiations between Israel and the Palestinians, important as they are, so far have only resulted in an interim agreement and the water situation for the Palestinians in the West Bank (and the Gaza Strip) is still one of the most sever worldwide and unequally more difficult than that of their co-riparians (Phillips et al., 2004; UNDP, 2006).

However, aside the difficult situation and the persistent problems, Jägerskog (2003) nevertheless concludes that the ongoing process of water collaboration in the Lower Jordan River Basin has proved to be rather robust and resilient. This is demonstrated by the facts that the water cooperation was continuing between Israel and the Palestinians despite of the Second Intifada in 2000 and furthermore, Jordan and Israel managed to solve problems that have surfaced, for example, when there was a dispute over allocations during the drought in 1999, through communication and dialogue (Jägerskog, 2003).

2.1.4. Institutional Setting in the Jordanian Water Sector

Planning and regulation in the light of limited water resources has ever been a central political issue in the Lower Jordan Valley. With regard to the focus area of this thesis the following section presents a short overview on the development and actual status of the Jordanian water sector. For reviews of the institutional settings in Israel and Palestine please refer to the literature (e.g. (Fischhendler & Heikkila, 2010; van der Molen et al., 2011).

The timeline in **Fig. 2.2** portrays the historical development of the Jordanian water sector since the sovereignty of the Hashemite Kingdom which spans from an initial phase of water-service self-organisation over a period of institutionalisation and centralization and towards the recent strategy of IWRM-orientation and private sector participation.

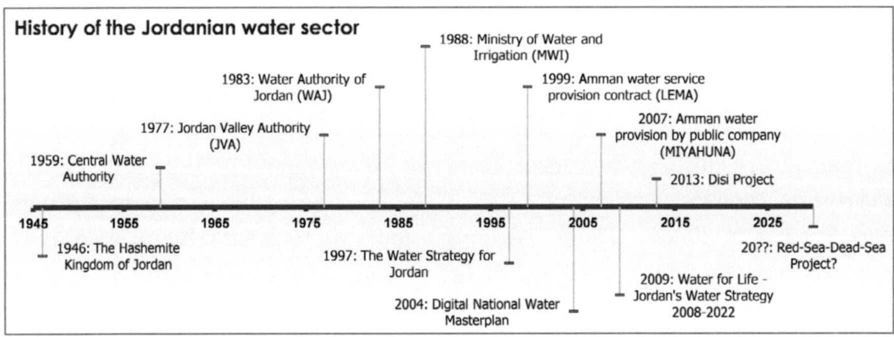

Fig. 2.2: History of key steps in the development of the Jordanian water sector since the sovereignty of the Hashemite Kingdom of Jordan.

Special importance in this regard has to be accounted for the major water sector reform during the late 1980s and the establishment of the Ministry of Water and Irrigation (MWI) in 1988 as the formally highest political decision making institution, thus dividing policy making (MWI) from service provision (Water Authority of Jordan (WAJ) and Jordan Valley Authority (JVA)). Since its establishment, the MWI has also been continuously supported by several donor organizations that have assisted in the development of water policy and water master planning as well as the restructuring of the water sector. On this basis and in the wake of a tight water situation after a series of droughts and in harmony with the international development of sustainability thinking, Jordan started to adapt integrated management principles in its national water policy during the 1990s. The first comprehensive national Water Strategy, adopted in 1997, consequently stressed the need for improved resource management with particular emphasis being placed on the sustainability of present and future uses (MWI, 1997). The 1997 Water Strategy embraced key aspects of resource protection, efficiency and equity of distribution, resource reuse and the use of unconventional water resources as well as a dual demand and supply management approach and can thus be considered as first Jordanian IWRM policy instrument. Another fundamental step towards integrated management in Jordan has been the launch of the digital National Water Master Plan in 2004 as a dynamic and regularly updated cross-institutional source of authorized water sector status and planning information (MWI, 2004). With the water supply management contracts in Amman and Aqaba, the MWI also began to endorse the expansion of private sector participation in water service provision. Various forms of operational arrangements or, more recently and on the basis of the gained experience, the establishment of publicly owned corporations have therefor been launched (Ditzel, 2008).

Although quite comprehensive in considered topics, the 1997 Water Strategy, however, remained fairly reluctant in directly demanding substantial changes in important aspects like for example the overall reduction of agricultural freshwater consumption or the full implementation of TWW reuse schemes. Therefore it was experienced to provide yet not to sufficiently stimulate the implementation of water laws, regulations and policies (Saidam & Ibrahim, 2006). Furthermore, implemented projects within the framework of this strategy that mainly emphasized the need to tap the full potential of available resource, have been experienced to be disadvantageous to the environment and additionally contribute to the depletion of the Dead Sea (Salameh, 2008).

Hence, the need to address these challenges and speed up the implementation process led to the preparation of the new water policy: *"Water for Life, Jordan's Water Strategy 2008-2022"*, which was adopted in 2009 by the Government of Jordan and includes a strategic water sector investment program and an implementation action plan (RCW & MWI, 2009). In contrary to the former strategy from 1997 this new policy demonstrates much more dedication to critical changes in water allocation and economic principles and also states concrete actions to be taken in these regards.

2.2. Integrated Water Resources Management

The importance as well as the challenges of achieving future water security has already rooted in the global awareness nowadays (cf. Chap. 1.1). The right way forward, however, remains item of research and dispute and for some part even belief.

In this respect, the Integrated **Water Resources Management** (IWRM)[3] approach has been the most prominent strategic water management concept in the international discourse during the last decade. The intellectual evolution of the IWRM idea has to be understood in different facets and steps. Although it appears difficult to define the exact origin of the term IWRM itself, the United Nations Water Conference in 1977 in Mar del Plata is often seen as fundamental benchmark in the development of a global water awareness (Biswas, 2004). The contemporary rising international recognition, however, certainly started from the Rio and Dublin Conferences in 1992 in accordance with the general strengthening of sustainable development ideas and international integrated resources management thinking.

However, coordinated efforts towards integrated management in the form of comprehensive and strategic planning practices have already been undertaken for a considerably longer period. The basic moving spirit hereto was the transition from the historical approach of sectoral separated towards a holistic planning and implementation strategy of water issues and related resources. This transition originated in the wish for an improved coordination and development between competing uses (e.g. flood and pollution control, water supply and conservation) under increasing usage pressures resulting from population growth, intensified irrigation and industrialization. This progress already started at least during the first half of the 20th century with the multiple purpose river development practices in the USA in the 1930s and eventually on international level from the 1950s (Mukhtarov, 2008). Rahaman & Varis (2005) even recognize century old forerunners of some basic IWRM principles in some regions of Spain where multi-stakeholder and participatory water tribunals have operated at least since the tenth Century.

---

[3] The exact term IWRM is used primarily in the European vocabulary. Labels of similar or comparable approaches in other regions are e.g. Integrated (Water) Resources Planning (IRP), Total Water Management (TWM), Integrated River Basin Management (IRBM), Shared Vision Planning (SVP) and others.

Besides this sectoral coordination, it has especially been the spatial integration that received early attention in water management. There are approximately 261 international river basins today (Wolf et al., 1999) and good water relations have always been a factor of political stability between co-riparian administrations. While historical transboundary cooperation mostly concerned issues of navigation, borders and single purpose rights, it has developed in accordance with the water sector to nowadays serve multi-purpose basin strategies. Hamner and Wolf (1998) listed 145 transboundary signed treaties dating between 1870 and 1984 that specifically negotiate water management matters between riparian nations (excluding navigational, boundary and fishing rights).

Following this tradition of trans-sectoral and transboundary integration and coordination the contemporary IWRM understanding is furthermore clearly guided by the principles formulated by the International Conference on Water and the Environment in Dublin in 1992:

1. Freshwater is a finite and vulnerable resource, essential to sustain life, development and the environment.
2. Water development and management should be based on a participatory approach, involving users, planners and policymakers at all levels.
3. Women play a central part in the provision, management and safeguarding of water.
4. Water has an economic value in all its competing uses and should be recognized as an economic good.

Although criticized for their generality and especially for the heavily disputed aspect of water commodification (Barlow & Clarke, 2002; Biswas, 2004), these basic principles have been regularly confirmed in subsequent international forums during the last two decades (for an interesting overview hereto see for example Rahaman & Varis (2005)). Moreover, additional aspects also achieved recognition as fundamental IWRM topics, including facets of social equity and poverty alleviation in water services, the role of water as a patron of peace or conflict, the challenges of imminent global changes, the appropriate levels of implementation or the necessity of highly adaptive strategies in order to realize any of the envisioned goals. And as the extensive body of available literature and active research suggests, the journey towards a comprehensive understanding of the requirements, boundaries and possibilities of truly holistic and sustainable water sector strategies appears far from ending any time soon.

In view of this continuous development process as well as due to the universal and visionary character of the IWRM principles it is not surprising that various definitions have evolved that try to grasp the concept. A comparison of some major players in the IWRM domain highlights some interesting commonalities and differences in the IWRM understanding (**Table 2.2**).

Table 2.2: Selection of definitions used by some major actors in the IWRM domain.

Organization (Reference)	"IWRM…"
Global Water Partnership (GWP, 2000)	*"…is a process which promotes the coordinated development and management of water, land and related resources, in order to maximize the resultant economic and social welfare in an equitable manner without compromising the sustainability of vital ecosystems"*
USAID (USAID, 2002)	*"…Is a participatory planning and implementation process, based on sound science, which brings together stakeholders to determine how to meet societies long-term needs for water and coastal resources while maintaining essential ecological services and economic benefits."*
U.S. Army Corps of Engineers (USACE) (Cardwell et al., 2006)	*"…is a goal-directed process for controlling the development and use of river, lake, ocean, wetland, and other water assets in ways that integrate and balance stakeholder interests, objectives, and desired outcomes across levels of governance and water sectors for the sustainable use of the earth's resources."*
The World Bank (Millington et al., 2006)	*"…aims to establish a framework for coordination whereby all administrations and stakeholders involved in river basin planning and management can come together to develop an agreed set of policies and strategies such that a balanced and acceptable approach to land, water, and natural resource management can be achieved."*

Perhaps the most frequently quoted definition has been developed by the Global Water Partnership (GWP), who has also long been a major advocate of the IWRM concept. It explicitly states the need to jointly address water, land use and other resources issues, thus requesting a very holistic management approach reaching beyond the water sector alone. In this point the cited World Bank definition agrees but at the same time it puts more weight on the aspect of IWRM as participatory process with a need for stakeholder involvement. Participation appears also the leading aspect in the USAID understanding, the latter also giving the sole definition directly demanding an integration of scientific views and approaches. The aspect of economic, social and environmental consideration in the water management process, appears to be inherent in all discussed definitions, however, it is most emphasized in the wording of the GWP and the USAID. On the other hand, the World Bank definition and to a lesser extent the

USACE, appear to shift from the archetypical goal of sustainability to rather promote a "balanced and acceptable" (or in other words: less un-sustainable) utilization of the resources. It is furthermore interesting to note, that GWP, USAID and USACE understand IWRM essentially as a process, not a state or a concept, thus implicating continuity. The World Bank definition takes an exceptional position here as well, describing IWRM as a coordinative framework to be established as a platform for balanced water management processes.

Although helpful for a first contemplation of the subject, such musings on definitions, however, only succeed to scratch on the surface of the IWRM matter; and it has to be kept in mind that all such short definitions carry along an almost overwhelming volume of discussions and considerations in the form of reports, background papers and recommended guidelines. Moreover, defining IWRM within one or two sentences appears to advocate the approach as fully sensible and pragmatic procedure, and consequently as the rationally favourable path towards sound and responsible water management. This, however, is not undisputed, and the complexity of the subject becomes clearer when one considers the multiplicity of processes necessary to implement the stated principles as well as the vast amount of tools that have been recommended for their facilitation (**Fig. 2.3**). The insight that immediately follows such examination is twofold:

First, the theoretical construct of IWRM cannot be understood as operational guideline and any local or regional initiative eventually has to develop its own specific practice adapted to the given situation (GWP, 2000).

Second, attempting to fully implement the theoretical IWRM logic means that before any truly integrated management decision can be conducted, so many prerequisites of finding right balances and levels of integration have to be fulfilled that the whole concept appears hopelessly ideological (Lankford et al., 2007).

This is also where the main critiques of the IWRM approach set in. Biswas (2004), for example, sees a fundamental flaw of the whole concept resting in the vagueness of its core principles and the resulting lack of any operational guideline, a case he states especially for the concerns of developing countries. Other authors are generally less fundamental in their critique but agree that, despite the extraordinary investments and efforts channelled towards IWRM during the last decades, the theory has not yet been able to provide sufficient empirical evidence for its benefits (e.g. Galaz, 2007; Jeffrey & Gearey, 2006; Merrey, 2008). And due to the mentioned lack of guidance, Molle (2008) as well as Merrey (2008) fear that water managers in their struggle to implement the full normative IWRM package, are lead to a paralysis rather than to a prioritized set of actions. Or as Merry puts it, the concept that set out "*...to help water* *professionals 'think outside the box' [...] is now becoming a new box confining thinking and action."* (Merrey, 2008).

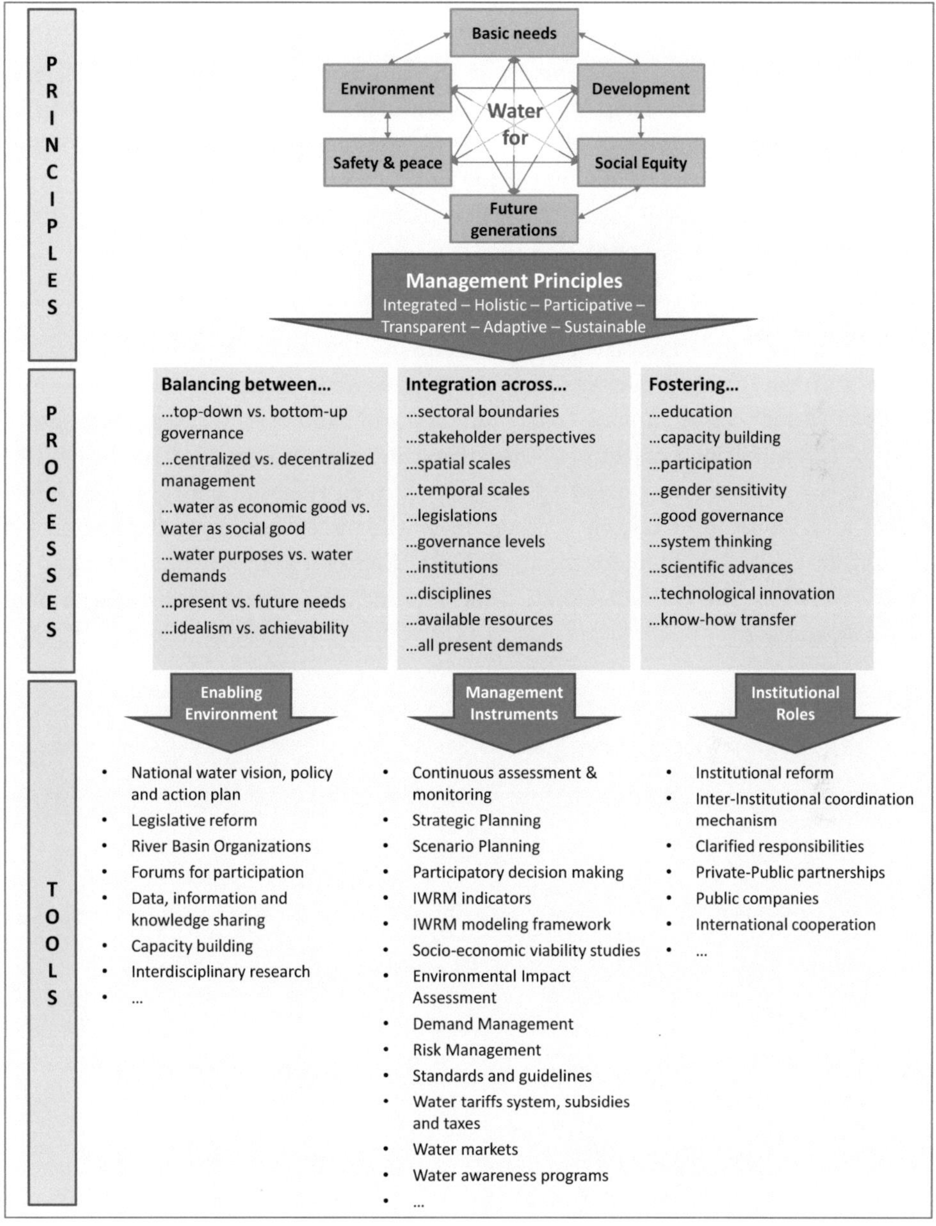

Fig. 2.3: Principles-Processes-Tools in the IWRM framework.

These critical reflections are perhaps in large parts rooted in a frustration about the very slow or sometimes not even visible progress of societies towards the ambitious visions of the Millennium Development Goals. And of course they appear valid and very important in generating the insight that the current value of IWRM thinking lies more at the normative and strategic level, hence describing what best should be, and not so much at the implementation level, guiding how things are best done (Mitchell, 2004). Thus, the IWRM process perhaps has to be understood as much a process of learning to implement as implementing itself. Some authors have already labelled these notions, among some other conceptual revisions, under the new term "Adaptive Water Management" (Gleick, 2003; Medema, 2008; Pahl-Wostl et al., 2005).

Although many fundamental questions remain unanswered, it is nonetheless obvious that the last two decades have been witness to tremendous changes and reforms in water sectors in many parts of the world. And even if not yet successful in any of the ambitious objectives, the strong urge for integration in the IWRM paradigm has at least succeeded in bringing water professionals together, inter-sectoral and transboundary, in a community that goes far beyond questions of legitimation (*see also*: van der Zaag (2005)). Especially in an environment as conflict-prone as the Middle East, this already has to be perceived as an invaluable success.

2.3.   Decision Support

Rational decision making is obviously a problem of choice between alternative ways of action with the aim to find the best possible (or most appropriate) solution according to given objectives. When the decision objectives are based on more than one criterion, which usually is the case, the decision problem becomes complex and requires a systematic approach, presumed a rational choice is desired. Once multiple objectives and criteria as well as uncertainties about the different alternatives pay-offs come into play, the decision situation becomes ill-structured, which basically means that the choice cannot be automated but requires a true cognitive decision. Typical questions the decision makers are confronted with in such situations are related to the benefits, conflicts and efficiency of alternative solutions with regard to their defined objectives. In this sense Roy (1996) understands decision support (also: *decision aid*) as

"[...] any activity which, by explicit but not necessarily fully formalized models, helps to obtain elements of responses to the questions posed by a participant in a decision process" (Roy, 1996).

Following the principles of IWRM, planning and decision making has to be collectively evaluated from various viewpoints, corresponding to the knowledge of experts in different domains as well as to the interests of various stakeholders. Furthermore, the decision situations addressed are usually highly ill-structured or often even "wicked", the latter describing problems that defy even an unambiguously correct formulation (Rittel & Webber, 1973). In such a context each person involved in the problem solving may see the problem from his own perspective and uses his own position to its characterization (Mysiak et al., 2005). In this regard any IWRM approach conveys the urgent requirement for efficient and transparent decision support, which is also recurrently stated by scientists and practitioners alike (GWP, 2000; Kok de & Wind, 2003; Mysiak et al., 2005; OECD, 2006).

The comprehensive nature of the IWRM process hereby recommends different supportive strategies dedicated to different stages of the planning and decision making process. An exhaustive collection of the countless promoted strategies cannot be given here, but the following sections are meant to provide an overview through a short discussion of some exemplary and popular approaches.

2.3.1.   IWRM Planning Frameworks

On a very generic scale, the process of planning and decision making is classically divided into four major phases: intelligence, design, choice and implementation, which again can be subdivided to identify several key steps to be addressed during the process. (Simon, 1960; Sprague & Carlson, 1982). In the effort to achieve good practice in integrated water resources planning, many authors have already adapted this perception and proposed flowchart-like guidelines for a stepwise organization of the process (**Fig. 2.4**) (Becker & Hattermann, 2005; Liu et al., 2008; Malczewski, 1999).

The understanding of the actual state and the urgent challenges of the observed system is perceived as functional starting point in any environmental planning project. This first "intelligence phase" comprises the formulation and identification of specific focus problems, particular objectives and possible response options. Thorough review of the available information and strong stakeholder participation are two key necessities in this stage of the decision process. Thus, efficient information retrieval procedures as well as techniques for structuring, visualizing and sharing the relevant and processed information provide significant help in this phase.

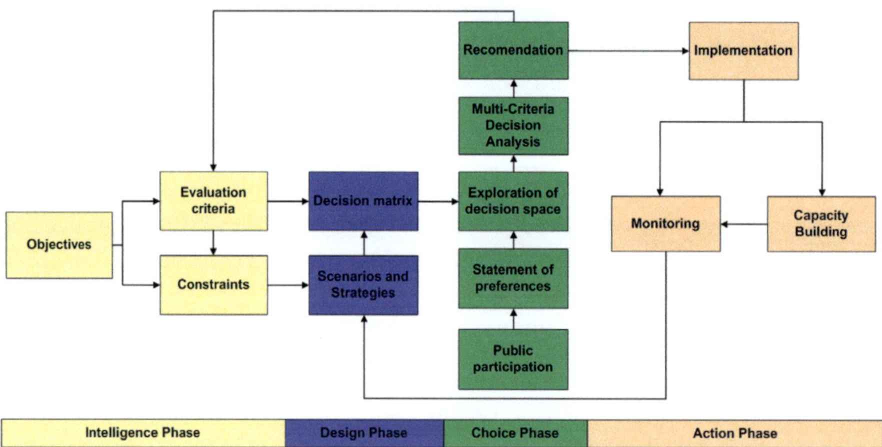

Fig. 2.4: Phases and steps in the IWRM modelling, decision and implementation process (modified after Malczewski (1999)).

2.3.2. Indicator Frameworks

Besides setting the objectives, the crucial outcome of this first phase is a clear formulation of the eventual decision evaluation criteria, or in other words the questions that will subsequently define the modelling requirements and finally govern the choice between the alternative options. In this context the need for suitable indicators arises, that allow to measure the performance of the different planning alternatives with regard to the stated objectives.

Recommended frameworks for the selection or the development of sustainability indicators, as well as proposed collections of indicators can be found in abundance in the literature. Especially the advent of environmental assessment and state of the environment reporting during the last decades have led several large institutions to propose own core sets of environmental sustainability indicators, e.g. the European Environmental Agency (EEA, 2005), the Organisation for Economic Co-operation and Development (OECD, 1993), the United Nations Commission of Sustainable Development (UN, 2001) and the World Bank (Segnestam, 2002).

The overarching intention of economical, ecological and social sustainability hereby results in some recurrence of advocated general indicator themes between different IWRM initiatives, such as for example the quantity and reliability of water supply, the provision, cost and affordability of supply and sanitation services, or the recognition of environmental water requirements. This has also led various authors to provide more or less exhaustive collections of indicators proposed as suitable to evaluate sustainability in general or to a certain topics therein (e.g. Esty et al., 2005; Garfi & Ferrer-Marti, 2011; Giannini & Giupponi, 2011). On the other hand, the particularity of each planning context inevitably leads to the necessity of specific indicator collections to account for the concrete objectives in adequate detail. Consequently, the number of IWRM indicator sets that have been proposed probably comes close to the total of IWRM initiatives that have been conducted.

A popular approach to a systematic development of such indicator sets is the **Driver-Pressure-State-Impact-Response** model (DPSIR). Introduced by the European Environment Agency (Smeets & Weterings, 1999) as *"a framework for describing the relationships between the origins and consequences of environmental problems"*, it was basically an extension of the former Pressure-State-Response (PSR) framework used for the State of the Environment reporting of the Organization for Economic Cooperation and Development (OECD, 1993), which itself was derived from the stress-response approach of Rapport and Friend (1979).

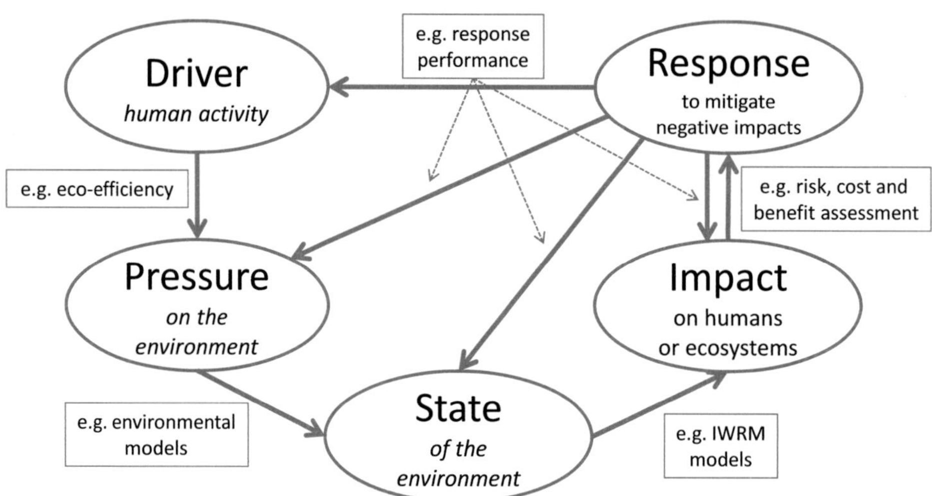

Fig. 2.5: Framework of the DPSIR-model. The rectangles propose possible measures and/or indicator categories applicable in the different stages (modified after Smeets & Weterings (1999)).

The DPSIR framework is meant to provide a structure in which to identify and organize environmental indicators. Therefore it uses a model of cause-effect relationships that link human activities to their eventual environmental impacts and to the consequential societal response options. As depicted in **Fig. 2.5** the chain of causal links starts with human activities as 'driving forces' (e.g. an agricultural irrigation practice) that create 'pressures' on the environment (e.g. groundwater abstraction rates) which result in a change in environmental 'states' (e.g. sinking groundwater tables) that might have 'impacts' on other human needs, health or ecosystem functions (e.g. drying out of nearby drinking water wells) and eventually require political and societal 'responses' which might aim to address any of the former stages (e.g. irrigation regulation or water imports).

The DPSIR concept is used as indicator framework for the State of the Environment Reports in the European countries as well as in the United Nations Environment Programme (UNEP), where it is especially promoted as a tool to *"enable feedback to policy makers on environmental quality and on the resulting impacts of the political choices made, or to be made in the future"* (Kristensen, 2004). It also has been adopted by Decision Support Systems in the field of IWRM (e.g. MULINO, WaterStrategyMan).

It is important to understand that an indicator set in the DPSIR framework should not be misinterpreted as a comprehensive model of environment-socio-economy-relations, since the complex system is necessarily broken down to rather simple causal chains. In this respect the DPSIR concept also provokes some critique as it often appears difficulties to draw clear boundaries between the categories and a certain indicators might thus be assigned to different types. For instance, a type of land cover can be understood as a *State* indicator, being dependant on different *Driving forces* of human activity, or it can be seen as a *Driving force* itself, influencing other *State* indicators, e.g. soil properties or surface runoff (Moxey et al., 1998). The most serious objection to this approach is that it neglects the systemic and dynamic nature of the processes, and their embedding in a larger total system containing many feedback loops and multiple pressure and impact relations (Bossel, 1999; Müller & Wiggering, 2004).

Besides the DPSIR model, a broad range of other indicator frameworks can be found in the realm of sustainable development studies (e.g. Giannini & Giupponi, 2011; Niemeijer & de Groot, 2008) and the theoretic literature on indicators is quite extensive. Additional information on the topic can be found for example in Helming et al. (2008).

2.3.3. Decision Support Systems

Transferring the broad definition of decision support as given in the introductory section of this chapter (cf. Chap. 2.3) to a technical level, means that decision support is already given by systematic data organization as provided by any information system as well as through the analytical capabilities of respective domain models. And decision support systems (DSS) hence are commonly understood as:

"[...] interactive computer-based systems that help decision makers to to retrieve, summarize and analyze decision-relevant data." (Power, 2002).

Systems designed to that purpose have been developed since the 1970s from the basis of rather simple information systems towards systems that were capable of assisting in problem analyses. Hereby, the main task assigned to the DSS was always to assist the decision maker in the exploration of large amounts of information and its reduction to a few core indicators to be evaluated for the alternative solution options (Gupta & Harris, 1989). Especially in the fields of operations research in industry, engineering and management this led to the development of the theoretical basis of technical decision support and decision support systems (e.g. Blanning, 1979; Geoffrion, 1983; House, 1983; Mittra, 1986; Sprague & Watson, 1986)

Concerning the schematic architecture, many authors (e.g. Haag & Kaupenjohann, 2001; Power, 2002; Sprague & Carlson, 1982 and others) identify the three major components of DSS as: (a) Database and Database Management System, (b) Models and analytical tools and a related Model Management System, (c) A user interface which enables an interactive dialogue between decision maker and DSS. Haettenschwiler (1999) and Marakas (2003), among others add another important component to this conceptual architecture: (d) the User. When the system, through incorporation of a GIS component, is furthermore capable of spatial data analysis it is often referred to as spatial decision support system (SDSS) (Batty & Xie, 1994; Fedra et al., 1992).

The above cited system architectures are very much focussed on the information system part of a DSS. But the actual process of finding a final choice between alternatives in a complex and ill-structured decision situation also allows for formalization and support within decision models. The typical approach hereby is for the decision maker(s) to assign weights to the preliminary defined goal criteria in accordance to their preferences and to afterwards rank the performance of the decision alternatives with respect to the goals and given weights. Various systematic approaches have been proposed thereto that are commonly subsumed under the domain of Multi-Criteria-Decision-Analyses (MCDA; MCA). Typical MCA-models are *Outranking* approaches (e.g. ELECTRE (Roy, 1968) or PROMETHEE (Brans & Mareschal, 1994), *Pairwise comparisons* (e.g. AHP (Saaty, 1987)) or *Fuzzy MCDA* (e.g. Fuzzy set analysis (Buckley, 1985)).

The complexity of the IWRM process has always challenged researchers and practitioners to develop dedicated decision support models and tools. Some address rather technical questions to support engineers, scientists and planning managers, others aim to provide support for policy decisions by administrators and public participants from the stakeholder groups. Consequently there have been numerous DSS-labelled approaches developed within the water resources management field that can be loosely classified as:

Specialised domain models are probably the most straightforward transfer of scientific understanding into decision support. These models are usually very specialized and often require considerable expertise in their handling. Typical examples of this category in the IWRM domain comprise hydrological and groundwater models, technical and economic feasibility models (e.g. for decentralized waste water treatment plants) or operational models (e.g. for network maintenance or irrigation scheduling). By allowing the assessment of the impacts of decisions on the modelled system they build a robust basis of a decision support framework. Taken isolated, however, such specialized models cannot address the integrative needs of an IWRM planning situation.

IWRM models were therefore developed to provide an integrative platform to address the different subsystems of the IWRM domain. In most cases they emerged as river basin simulation models from a hydrological basis with a focus on water allocation and water balancing. Typical application objectives are to manage river basins operation and development, address conflicts in water uses or evaluate socio-economic and environmental impacts of alternative management strategies. A classical distinction is made between simulation and optimization models, although state-of-the-art models often contain elements of both (Wurbs, 2005). The general approach is to simulate the movement of water through a system of river reaches and nodes that represent reservoirs, diversions and abstractions, demand sites and other network elements, in order to simulate and optimize different allocation scenarios. For this purpose most applications adopt some form of Linear Programming solvers, but other optimization algorithms have also been proposed (e.g. dynamic programming, gradient search, genetic algorithms and others). The river basin model often provide a more or less interactive linkage to other model components, e.g. above mentioned specialized domain models, in order to fulfil the objective of IWRM modelling. Hundreds of IWRM-River basin models can be found in the published literature, of which prominent and elaborate examples are the models developed by the Hydrologic Engineering Center (HEC) of the USACE (HEC-HMS, HEC-RAS, HEC-ResSim), the MIKE model family (MIKE SHE, MIKE BASIN) from the Danish Hydraulic Institute (DHI), the MODSIM model (Labadie, 2005), RiverWare (Zagona et al., 2001), RIBASIM by Delft Hydraulics as well as the WEAP model by the Stockholm Environmental Institue (SEI) (Yates et al., 2005).

Participatory decision making tools are focussed on supporting the process of finding a consensus and an eventual choice between different alternative ways of action. In this sense they are probably closest to match the actual term decision support. Proposed software tools typically employ models from the set of Multi-Criteria-Decision-Analyses techniques and are not necessarily specialized on water management issues alone but decision situations in general. Special focus within the integrated water resources management domain (but also not exclusively there) has, however, been given to the support of participatory decision making. Depending on the technique and decision model applied the process can focus on finding the most rational choice (optimization models) or one that applies best to the perspectives of the decision makers and stakeholders (Weighing and Ranking models). An exemplary tool for participatory decision making in the IWRM field was developed within the MULINO (MULtisectoral, INtegrated and Operational Decision Support System for Sustainable Use of Water Resources at the Catchment Scale) framework (Giupponi, 2007). A very helpful review on further types and applications of MCA in the field of water resources management is given by Hajkowicz & Collins (2007).

2.3.4. Scenario Planning

Scenario Planning can basically be understood as the art of alternative thinking in long-range strategic decision making processes (Godet, 2000).

When we imagine the future we usually rely mostly on our memory of past experiences. Moreover, psychology and neurosciences provide evidence that the cognitive processes of recalling empirical knowledge and of planning for the future basically utilize the same neural mechanisms and allow us a *"mental time travel"* into our past as well as into an imagined future (Tulving, 1985). In this regard Schacter et al. (2007) propose to think of the brain as a fundamentally prospective organ that is designed to use information from the past and the present to generate predictions about the future".

These considerations imply two important aspects of how we usually deal with the inherent uncertainties of future planning and decision making.

On the one hand, uncertain aspects of the future that are strongly linked with personal experiences are often naturally considered within what appears to us as a plausible range of expectations. For example, planning with variable weather conditions for a weekend trip to the mountains seems commonsense, as we probably have a solid empirical knowledge of the variability of mountain weather. So, in this example we might naturally plan our actions on the basis of a set of different plausible futures.

On the other hand, when asked to systematically construct a longer, yet still imaginable (e.g. 5-15 years), view into the future we typically lack this close personal memory of the variability and uncertainty of the different development aspects that have to be considered. And even if we are aware of the plausibility of variance because it had so been experienced in historical events, this understanding is usually far less strongly linked with experienced memory but more with theoretical knowledge. In such situations we tend to fall back to what appears to us as most rational, which is to anticipate a most likely *"official future"* by projecting current conditions and trends into the years to come (Schwartz, 1991). And actually there are many cases in which such trend predictions are experienced to prove reasonably accurate, e.g. in the case of demographic projections. The weak spot of the "official future" in the light of planning and decision making, however, is that it is often observed as single probable future, thus limiting the possible variability assessed in the planning view and resulting in rather concealing risks than revealing risks (van der Merwe, 2008). And especially when the planning exercise has to consider elements within quickly changing circumstances, such as for example in business or policy settings, history has shown that simple extrapolations of past trends, are unlikely to produce reliable forecasts in the medium or long term (Makridakis et al., 2009).

The concept of scenario planning (also: *scenario learning*) has been developed to offer a systematic approach to consider a range of plausible futures rather than to focus on the attempt of a highly accurate prediction of a single most likely outcome (**Fig. 2.6**). The principal focus hereby is put on the alternative development of the critical uncertainties of the observed system as well as on the possible occurrence of unexpected events that disrupt observed trends (also: *wildcards*). According to Schoemaker (1995) this involves a preliminary distinction between the aspects *"we think we know something about"* and which we perceive as better predictable and thus might leave to follow the momentum of their current trends, and those aspects *"we consider uncertain or unknowable"* and that will diverge within their plausible range and thereby define different future scenarios. Schoemaker (1995), Wack (1985) and others also emphasize that a set of scenarios must not be understood to represent a mutually exclusive and exhaustive set of future states but rather as an attempt to bound the futures inherent uncertainties. Thus, the important idea of scenario-building is not to make predictions or forecasts, but rather to provide images of the future that challenge prevalent assumptions and broaden perspectives.

The numerous scenario definitions found in the related literature basically follow the above considerations (e.g. (Fahey & Randall, 1998; Godet, 2000; Schoemaker,

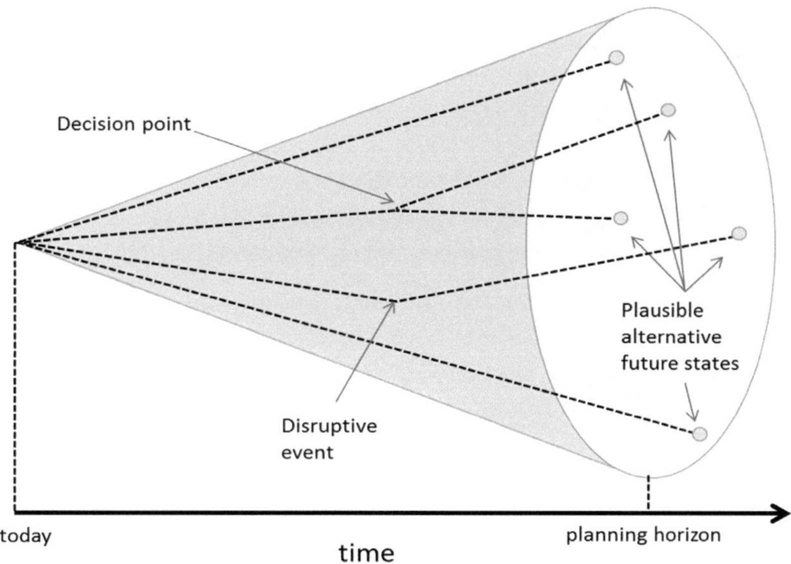

Fig. 2.6: Scenario Funnel (adapted from Timpe & Scheepers (2003) and Gausemeier et al. (1998))

1995; Schwartz, 1991; van der Heijden, 1996)). A suitable synthesis is found in the definition used by the Intergovernmental Panel on Climate Change (IPCC):

"A scenario is a plausible description of how the future may develop based on a set of coherent and internally consistent assumptions about driving forces and key relationships. Scenarios may be derived from projections, but are often based on additional information from other sources, sometimes combined with a narrative storyline."(IPCC, 2007).

The concept of thinking about alternative futures, utopias and dystopias is surely dating back to earliest philosophical considerations in human history. The formal development of modern day strategic scenario planning methodologies, however, appears to root in US military strategy exercises during World War II and their subsequent extension towards the fields of social forecasting and public policy in the early 1960s (Bradfield et al., 2005). However, it appears that for the most part of the last century, significant consideration to the concept had only been given by practitioners and not so much by academics. A growing number of companies started to adapt scenario planning for their long range strategic decision making, including for example Royal Dutch/Shell, IBM and General Motors, whereof especially Royal Dutch/Shell repeatedly reported great success stories (Chermack et al., 2001), which also secured them the prevalent pioneering role in the development and application of strategic planning scenarios. After an observable decline of interest in the commercial world during the 1980s (Chermack et al., 2001), scenario planning has experienced a vivid renaissance during the last two decades, also because it was discovered as a tool for interdisciplinary long-range policy analysis especially in the course of sustainable development planning (e.g. Anderson et al., 2003; Baker et al., 2004; IPCC, 2007). In this course the subject has also attracted increasing attention of the academic world in both, the natural science as well as the business study domains (e.g. Goodwin & Wright, 2001; Mahmoud et al., 2009; Mietzner & Reger, 2005; Nguyen et al., 2007).

The longer history of the scenario planning concept and especially its highly heterogeneous field of application have resulted in such a wide variety of proposed techniques that it already has been described as a "methodological chaos" (Martelli as cited in Bradfield et al. (2005)). A matter surely contributing thereto is the fact that scientific evidence for the functionality of a certain scenario methodology is difficult to obtain. Various authors have also criticized the lack of a clear terminology and a resulting ambiguity of used definitions to the point of "misuse and abuse" of the terms strategy and scenario (Godet, 2000). Consequently there have also been almost as many attempts on finding suitable categorizations of scenario types and methodologies, which, however, rather results in an additional classification confusion (for an overview please refer to Mietzner & Reger (2005)). An important aspect found as distinctive scenario characterization is their nature of being either exploratory or anticipatory (Godet, 2000), the former using forecasting methods to lead historical and present trends to possible futures, whereas the latter are built on the basis of alternative visions of the future and then construct sequences of developments that could lead there (backcasting). Other distinctions include quantitative versus qualitative scenarios, individual versus group scenarios, stakeholder versus expert scenarios or according to their originating methodological schools: (i) Intuitive Logics, (ii) La Prospective and (iii) Probabilistic Modified Trends (Bradfield et al., 2005).

Despite the apparent methodological jungle in the scenario planning domain, some common ground can still be identified as the majority of approaches still agree on a set of basic steps as originally developed in the works of Royal Dutch/Shell which were later also proposed in the foundational work of Schwartz (1991). On the same foundation Mahmoud et al. (2009) proposed a formal framework for scenario development in environmental decision making processes which puts a stronger focus on scientific modeling-based approaches. **Table 2.3** shows a comparative list of the sequential scenario processes as proposed by the mentioned authors. It is obvious that these steps have to be understood as of very generic nature and might each comprise a multitude of subroutines, the inclusion of various actors and different modeling as well as forecasting techniques.

Besides the scenario method described here, other techniques with the purpose to create, present, manipulate, and evaluate images about alternative futures have been elaborated. Notable methodologies include trend exploration, contingency planning, sensitivity analysis, the delphi method (Linstone et al., 1975) and morphological analysis (Ritchey, 2006; Zwicky, 1969).

Table 2.3: Comparison of the sequential scenario development phases according to the processes proposed by Schwartz (1991) and Mahmoud et al. (2009).

"Intuitive Logics" approach by Schwartz (1991)	Environmental model-based approach by Mahmoud et al. (2009)	
1. Definition of the focal topic or decision issue and the spatio-temporal scope	1. Scenario definition	
2. Identification of the key factors that govern the defined topic or influence the decision (decision criteria)		
3. Identification of the external driving forces that influence the development of the key factors (decision criteria)	2. Scenario construction	System conceptualization
4. Identify the critical uncertainties in the future development of the driving forces		
5. Definition of the basic scenario logic and the initial plots		
6. Full construction of scenarios by composition of plausible alternative development plots that might lead to the envisioned future state of the driving forces and simulation with adequate forecasting or backcasting techniques		Model selection or development
		Data collection and simulation
	3. Scenario analysis	
7. Assessment of the implications on the decision criteria, the objectives and strategies and the proposition of possible mitigation options.	4. Scenario assessment	
	5. Risk management	
8. Follow-up research and monitoring of "signposts" that indicate in which direction the actual development unfolds	6. Post-audit and monitoring	

2.3.5. Summary

To summarize the previous elaborations on decision support and related approaches it seems important to emphasize once more that the subject truly spans a much wider domain than could be discussed here. Moreover, in such an integrated domain as IWRM, complex decision processes take place during all planning and implementation steps and at various technical and policy levels. In this respect the presented methodologies are never to be understood as mutually exclusive and a holistic IWRM decision support framework would probably employ several or all of them as well as some other approaches that were not discussed here.

Furthermore, the careful reader might have noticed the terminological ambivalence occurring between (and even within) the fields of decision support, strategic planning and scenarios. Thus, it appears most sensible to conclude with a summary of the key terms and their definitions and interrelations as used in the further course of this thesis:

A **Decision** situation involves: **decision objectives** and, depending on the complexity and abstraction grade, further sub-objectives; a set of **decision alternatives** that might comprise single or multiple **potential options/actions** (e.g. Responses in the DPSIR model); a set of **decision criteria** used to evaluate the alternatives according to the objectives. In complex systems the criteria are often understood as **indicators** that describe the state of the system (Pressure, State and Impact in the DPSIR model). They are not required to be independent of each other but it is beneficial for a rational choice if they are not redundant; a **decision model** is the logic and reasoning that is used to ultimately choose between the decision alternatives according to the (formerly evaluated) criteria.

A **vision** is consent of a desired future condition that becomes tangible and distinct in a set of stated **goals**. For example the IWRM principles can be understood as a wide-ranging vision of equitable and sustainable water management with a set of stated goals related to the meaning of equity and economic and ecologic sustainability. The vision and the general goals become explicit in a set of **strategic objectives**.

A **strategy** is the long-term systematic approach (also: policy; roadmap) towards achieving the vision and goals by means of utilizing the available resources. In this regard a strategy incorporates the **strategic objectives** and the basic directional decisions that will consequently govern the subsequent operational decisions and actions within the **strategic planning** framework. As a strategy is dependent on the assumption of external as well as internal conditions and constraints (scenarios), it has to be regularly updated in the light of new developments. The **strategic plan** is a coherent and explicit set of operational decisions about the use of resources and actions to be undertaken to achieve the strategic objectives.

A **scenario** is a consistent and plausible description of a possible development pathway of a set of driving forces that leads towards a certain future state of the observed decision criteria at a defined scenario horizon (or planning horizon). It contains a scenario narrative as well as qualitative and quantitative information on the driving forces spatio-temporal development.

Driving forces or Drivers are the essential societal or environmental processes and trends that effect the development of the observed criteria. In the DPSIR model the drivers are usually assumed as human activities, whereas in scenario planning they are also understood environmental forces (e.g. climatic factors). In both models driving forces are assumed as difficult to directly influence by the decision maker and in this regard it is also sometimes differentiated between external and internal drivers, whereof the latter are at least in the sphere of possible influence.

2.4. Knowledge Management

2.4.1. What is Knowledge Management

First of all, "Knowledge Management" seems another one of those universally applicable, yet utterly vague terms that implicate a huge buzzword potential. In the case of Knowledge Management it becomes especially inexplicit, due to the trouble of grasping the term "knowledge" to begin with. And trying to generally reify the concept of knowledge inevitably leads into the thick of the realm of epistemology and a philosophical debate that is led at least since Aristotle's times.

In the wake of the information age, however, the recognition of knowledge as major intangible asset started to attract considerable attention to the topic in both academia and economics. Classically, there have been two groups of approaches towards Knowledge Management stemming from different origins. On the one hand a human-process oriented perspective, coming from the background of organizational learning and organization science and on the other hand a technical and structural approach from an information science and artificial intelligence background. Primary fields of interest have since been the relation of knowledge to information and modern information technologies, as well as the quest for efficient strategies to handle knowledge as a critical production factor in economic environments. The hence emergent multi-disciplinary field of Knowledge Management has proposed a large range of definitions for its subject. Put in the broadest context they can be summarized in the following characterisation:

Knowledge Management comprises the strategies and techniques that an organisation (or enterprise, or community) deliberately employs to create, gather, retain, organize, analyse, improve and share its knowledge in order to achieve its strategic objectives.

Given this definition it is important to note that Knowledge Management, while usually facilitated by technology, is not necessarily technology-bound. This highlights one major difference to the closely related field of Information Systems. However, in contrary to a philosophical and universal perspective, the knowledge definition in the Knowledge Management literature of course progresses from a more (information-) technical approach. And although it remains issue of considerable dispute within the discipline itself, to develop a working definition of knowledge is perceived as fundamental prerequisite to understand and undertake any knowledge management activity (Fahey & Prusak, 1998).

2.4.2. Defining Knowledge in Knowledge Management

A popular understanding of the concept of "knowledge" in the field of Knowledge Management is based on the data-information-knowledge-wisdom hierarchy (DIKW), often pictured in form of a pyramid (Ackoff, 1989; Rowley, 2007; Zeleny, 1987). The left flank of the pyramid in **Fig. 2.7** shows the "classical" DIKW hierarchy (the "Noise"-level is not included typically), which is grounded on the assumptions, that

- the concepts of data, information, knowledge and wisdom are distinct categories,
- there is a relation and a linear hierarchy between these concepts, and
- there is a kind of refinement of data via information and knowledge towards wisdom through some sort of filtering mechanism (hence the pyramid that narrows towards the top).

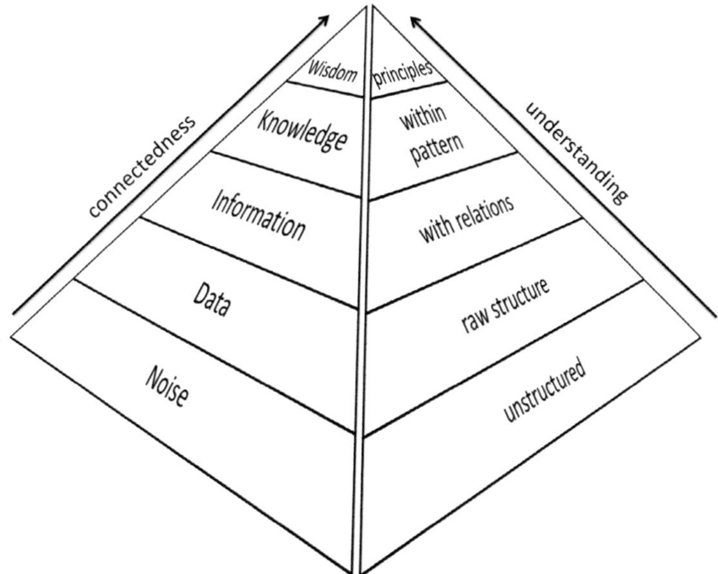

Fig. 2.7: The Data-Information-Knowledge-Wisdom-Pyramid (extended from Ackoff, 1989).

This simplified construction has been often disputed and modified in order to represent a higher complexity in the categories and their interrelations (e.g. (Matthews, 1997; Tuomi, 1999). Bellinger, Castro and Mills (2004) focused on possible transitions from data to information, knowledge and finally to wisdom, and found that it is the process of "understanding" that enables this transition from each stage to the next (cf. the right flank of the pyramid in Fig. 2.7).

More recent approaches recommend to completely abandon the pyramid image in favour of alternative models that account for the non-linearity of the data, information and knowledge concepts and their linkages (e.g. Faucher et al., 2008; Firestone & McElroy, 2004; Frické, 2009; Hicks et al., 2006). Nevertheless, even though debated for its reductionism, the DIKW model provides a useful basis for discussing some key differences between the concepts of data, information and knowledge and the eventual significance for understanding the terms Data Management, Information Management and Knowledge Management. Faucher et al. (2008) and Wallace (2007) both offer interesting overviews of different DIKW taxonomies from the Knowledge Management literature and conclude that the disparity of given definitions for the distinct categories rises from the data level towards the wisdom level. Though, in accordance to the hierarchical DIKW interpretation, most authors base their definitions of data, information, knowledge and wisdom on the respective concept below, which is "...processed in a meaningful way" to become part of the above level (Faucher et al., 2008).

Data is commonly understood as the unprocessed and elementary representation of reality as it is the product of mere observation. Often called raw facts or raw numbers, in the form of "...symbols that represent properties of objects, events and their environments." (Ackoff, 1989). For Bellinger, Castro and Mills (2004) these symbols "...simply exist and have no significance beyond its existence... [that] can exist in any form, usable or not... [and] does not have meaning of itself...". In contrary to this perception of complete absence of structure, it might be argued, that by the act of observing and recording data actually receives basic processing and structure (Tuomi, 1999).

The term **information** is already more vague but frequently understood as "processed data" that is organized in a specific context for a specific task (Drucker, 1995). According to Ackoff (1989) information is data, that is structured to answer "who", "what", "where", and "when" questions and thus supports decision making. And Bellinger, Castro and Mills (2004) state, that data becomes information by way of relational connection. The latter also implies that the distinction between data and information is primarily functional and not necessarily structural.

At the **knowledge** level, definitions eventually become highly ambiguous. In addition to Ackoff, who keeps to his reading and defines knowledge as processed information to answer "how" questions, Bellinger et al. (2004) understand it as a collection of information that means to be useful for some task. Others interpret the step from information to knowledge in that the latter one enables action (e.g.

Carlsson et al., 1996; Zack, 1998). In this sense, "knowledge" is directly connected and not separable from the "knowing mind" and the context of experience (e.g. Liebowitz & Wilcox (1997)). By general consensus on the basis of a comprehensive literature review, Alavi and Leidner (2001) conclude an information-technology centred definition for knowledge:

"Information is converted to knowledge once it is processed in the mind of individuals and knowledge becomes information once it is articulated and presented in the form of text, graphics, words or other symbolic forms." (Alavi & Leidner, 2001).

According to the mechanistic approach of artificial intelligence (AI), however, such a "knowing mind" theoretically does not have to be a human individual but can be formally represented as a deductive system of logical axioms, inference rules and heuristics. And in this sense knowledge is understood as:

A collection of facts, assumptions and relationships around a subject, as well as the rules that apply to this subject within the knowledge system.

From this AI perspective it seems now, that knowledge is explicitly expressible and separable from the knowing mind. Although in analogy to the precedent definition knowledge and knowledge system make no sense per se, but only in combination allow for interpretation, reasoning and decision making.

Taken together these two above definitions provide a useful working definition for knowledge in the author's perspective.

While data and information are usually considered as generic concepts, in the case of knowledge most authors tend to distinguish between different types, e.g. described as declarative ("know-what"), procedural ("know-how"), causal ("know-why"), conditional ("know-when") and relational ("know-who" or "know-with").

And another central distinction was found between the opposing types of tacit and explicit knowledge (Polanyi, 1966), suggesting that explicit knowledge is objective and can be easily expressed and codified into information (e.g. the stepwise process of a cooking recipe), whereas tacit knowledge is integral part of the human mind, highly subjective and cannot be recorded easily (e.g. speaking a language). This distinction has been in the focus of Nonaka and Takeuchi (1995) that modelled the possible transition processes between the two different types of tacit and explicit knowledge with the SECI-model (Socialisation, Externalisation, Combination, Internalisation).

Wisdom, despite being part of the classic DIKW hierarchy, is not often addressed in the Knowledge Management literature. Considering the above discussion of knowledge concepts, the "know-why" metaphor of Zeleny (1987) surely falls short. A more elaborate definition is given by Rowley (2007) who distinguishes wisdom from knowledge by adding a moral factor, which is: "The capacity to put into action the most appropriate behaviour, taking into account what is known (knowledge) and what does the most good (ethical and social considerations)."

To summarize the above discussion it can be concluded, that neither in the philosophical nor in the knowledge management debate, a distinctive definition of knowledge exists, that would indicate a straightforward application of a management procedure. There is, however, a uniform tenor on the existence of a relation between knowledge, information and data, in which the latter two are perceived as the vehicles of knowledge transfer. Thus, any knowledge management approach will have to elaborate its own understanding of the knowledge concepts and processes that it aims to support and how they relate to the data and information components as a means of knowledge transportation within its domain.

2.4.3. Knowledge Management Instruments

The plurality of views and perceived dimensions around the knowledge concept naturally brings a likewise wide variety of recommended strategies and instruments in the Knowledge Management landscape. Considering the above considerations, Knowledge Management tools thus comprise all instruments and techniques that allow themselves for the systematic support of an organizations knowledge process including the acquisition, transfer and utilisation of knowledge to fulfil the organizations objectives. In this respect, it must not necessarily be a matter of specialised instruments solely committed to KM tasks, but often is rather a matter of rather conventional technologies like databases, emails or discussion forums employed within a strategy to foster a successful KM environment.

And furthermore, a knowledge management strategy does not even inevitably involve the use of information technology, since a very important share of managing knowledge is considered to take place in organizational activities and routines, commonly referred to as the "knowledge culture" of an organization. Yet, KM approaches are usually technology enabled, today more than ever, and the following sections will continue to keep this technological focus.

In order to better understand the seemingly chaotic landscape of KM instruments promoted in the related literature, there have been numerous frameworks proposed to structure possible tools and strategies (Alavi & Leidner, 2001; Binney, 2001; Choi & Lee, 2003; Kühn & Abecker, 1997; Newman & Conrad, 2000; O'Dell et al., 1998; Roelof, 1999; Wang & Ahmed, 2005; Woods & Sheina, 1999). Many of such frameworks are based on or extending from the set of socially enacted knowledge processes as originally postulated by Holzner & Marx (1979):

- **Knowledge Acquisition** (also: Construction, Creation)
- **Knowledge Storage and Retrieval** (also: Storage and Organization)
- **Knowledge Dissemination** (also: Transfer, Distribution)
- **Knowledge Application**

Abecker & Kühn (1997) additionally emanate from considering the two major distinctive perceptions of knowledge, as they have also been discussed in the previous section (cf. Chap.2.4.2), and distinguish between what they call the "process-centred" and the "product-centred" approach. The process-centred view, as fostered mainly in the field of business and organizational KM, understands knowledge as an inevitably individual asset and sees KM primarily as a social communication process. Thus the role of a KM strategy is to enable, to facilitate, and to support communication and collaboration especially between individuals. The product-centred view, rooting in the fields of information science and AI, sees knowledge as a computable object and sets the focus on formalizing, storing and retrieving knowledge in a user-machine or even a machine-machine relation.

Based on these reflections **Fig. 2.8** illustrates an attempted classification of characteristic IT tools applied in the KM landscape.

Drawing close to the conclusion from the previous section about the non-uniqueness of the knowledge concept, the above discussion demonstrates that there cannot be a generic KM strategy, facilitated by the use of a specific instrument. A KM strategy must be adapted to the given objectives, the given environment and the people that eventually are both: the providing and receiving end of the knowledge. In this sense a holistic KM strategy is not facilitated by the use of a specific instrument, but rather must comprise a goal-oriented combination of techniques and tools that are respectively functional for a certain activity in the knowledge process. The borders here are of course highly fluid and very much dependant on the actual use case.

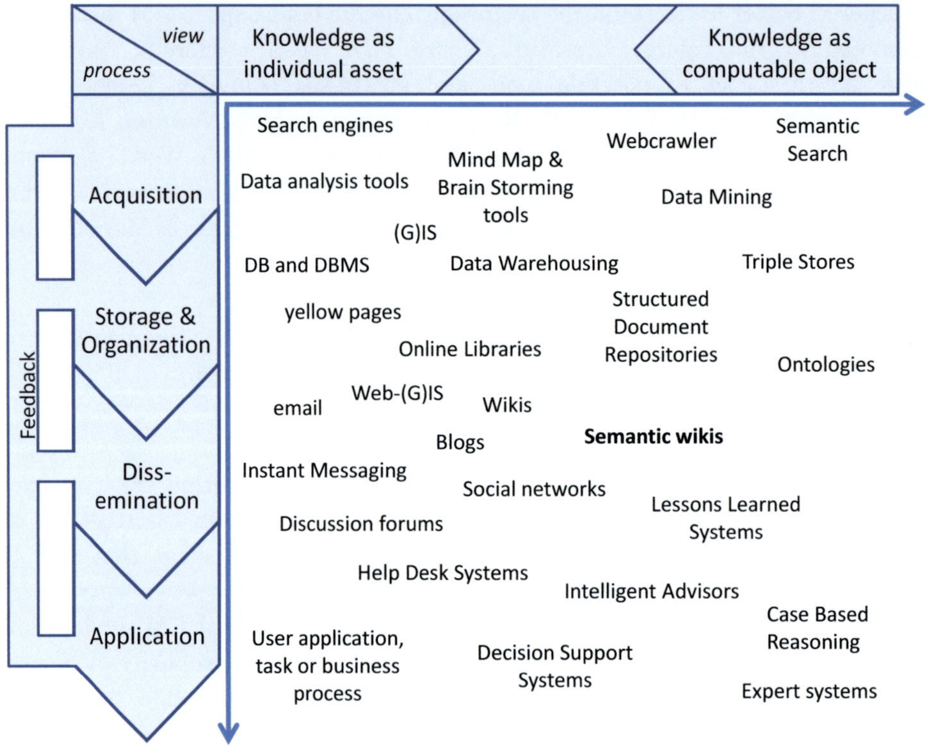

Fig. 2.8: Tools in the Knowledge Management landscape (extended and modified after Abecker (2004)).

Within the scope of this thesis and its given IWRM domain it was found that 'knowledge in need of managing' essentially relates to collaborative and comprehensive system understanding and the assumptions that are used to drive planning and decision making within such systems. Knowledge management therefore appears foremost as a means of facilitating the transportation and retrieval of such system understanding between participants in the process.

For this purpose the KM focus set in the presented work lies on one specific technology, semantic wiki, and its conceptualization and application within a use case of an Integrated Water Resources Management planning and decision support study. As stated above, one specific technology does not build an exhaustive KM framework but might be understood as one key element within a larger decision support framework. Semantic wikis are very much what the name suggests, a combination of semantic web and wiki technology. Both worlds will be briefly described in the next sections.

2.4.4. Semantic Knowledge Representation Structures

Externalization of knowledge requires formalization. The level of required formalization, however, varies considerably in dependence on the anticipated functionality and application of the knowledge representation. Brachman (1985) and later Guarino (1995) thereto distinguished between five levels of increasing formalization in knowledge representation: (1) linguistic, (2) conceptual, (3) ontological, (4) epistemological and (5) logical predicates. A computational treatment requires formalization at least on a conceptual level which includes a representation of relationships between expressed terms (Genesereth & Nilsson, 1988). On the ontological level these relationships are formally described themselves. But although expressive power that allows for automatic reasoning generally grows with logical formalism, Guarino (1995) argues that the beyond the ontological level the primitives used for description might become too general and thus too arbitrary for the task sharing and retrieval of knowledge. In agreement thereto, Schaffert et al. (2005) point out that currently the most successful knowledge models tend to be actually very simple and specific (e.g. Dublin Core, FOAF, as well as Thesauri like WordNet and DMOZ).

Thus, a suitable knowledge representation formalism would reside somewhere between the conceptual and the ontological level. A formalization mechanism for this purpose is often very generally called an ontology. According to the often cited definition given by Gruber (1993):

"An ontology is a specification of a conceptualization. That is, an ontology is a description of the concepts and relationships that can exist for an agent or a community of agents [...] and it is certainly a different sense of the word than its use in philosophy." (Gruber, 1993)

which was refined little later by Studer et al. (1998):

"An ontology is a formal, explicit specification of a shared conceptualisation." Studer et al. (1998)

In other words, an ontology in information technology, basically is a formal consensus between participants in the knowledge management process on the use of their vocabulary and the inherent interrelations. The practical use of such a shared vocabulary is that it permits unambiguous and consistent communication about a domain of discourse between human and machine agents, and thus allows to ask queries and to make assertions (Gruber, 1995).

In order to describe their domain, ontologies typically use clear identifiers (nouns) to state the existence of concepts (e.g. Municipality; Catchment; ...) and a subject-predicate-object-grammar to describe the relationships between them (e.g. Municipality *is_located_in* Catchment).

Emanating from this broad definition, ontologies themselves are can be understood in a formalization hierarchy depending on the potential of the applied grammar. Classification structures typically associated in this respect comprise (after Garshol (2004)):

Controlled vocabulary: An indexed list of explicit terms to be used (e.g. a list of fixed keywords) to describe a topic. Does only define the nouns but no relations between them and thus cannot be classified as an ontology after above definition.

Taxonomy: In the knowledge management literature used for a controlled vocabulary that is organized into a parent-child hierarchy. Restricts the relationship statements to one specific sort of generalization/specialization predicate (e.g. "is a", "is part of", "is instance of"). If the hierarchical link is formally defined, a taxonomy can be understood as a very lightweight ontology of weak expressiveness.

Thesaurus: Extends the set of possible predicates to express associations (e.g. "related to", "synonymous to"). Thereby, thesauri (again given the relations are well defined) are more powerful simple ontologies for knowledge structuring and retrieval.

Formal Ontology: Employs a formal representation language to define an own logic based subject description grammar (nouns and predicates as well as their specifications and constraints) to freely model the domain of discourse. Various formal ontology representation languages have been proposed which usually descend from first-order-logic or description-logic. Prominent examples are Resource Description Framework (RDF) (Lassila & Swick, 1999) and Web Ontology Language (OWL) (Motik et al., 2009) that have found widespread use in the semantic web environment. The goal of a well-defined ontology is to allow for automated logic reasoning (deriving facts that are not explicitly expressed) and thus for efficient information processing, structuring, retrieval and application. With this potential, formal ontologies are perceived as important enabler of the semantic web vision (Berners-Lee et al., 2001; Shadbolt et al., 2006). It is furthermore common to differentiate between top-level (also: upper) and domain ontologies. Top-level ontologies are generally attempting a very accurate description of fundamental axiomatic concepts and relations, such as space, time, matter, event, action, etc., to permit communication across domains. Domain ontologies ideally descend from these fundamentals to formalize their specific domain of interest (e.g. the infamous Pizza-Ontology).

2.4.5. Knowledge Management with Wikis and Semantic Wikis

Wikis (derived from the Hawaiian word for "quick"), and especially semantic wikis, can be considered as knowledge management instruments of the younger generation. Regarding their basic purpose as online collaboration tools wikis are often associated to the web 2.0 (the web of user generated content) development. Wikis enable online interaction and collaboration in the form of web pages which can be edited and structured with a simple mark-up language by anyone who has appropriate access and an internet browser.

The first wiki (called "WikiWikiWeb") was developed in 1995 by Ward Cunningham (Leuf & Cunningham, 2008), but the true success story began in 2001 when the internet encyclopaedia Wikipedia went online. And still today Wikipedia certainly is the most popular wiki application and has during the last years constantly resided in the top ten of the worlds most visited websites (Alexa, 2012). The wiki-technology itself is also applied in various other locations ranging from specialized *"Wikipedia-like"* encyclopaedias, over organizational knowledge bases and up to personal knowledge management applications. WikiMatrix lists 140 different wiki-engines (wiki software packages), and there are probably even more (Wikimatrix, 2012). The actual number of wiki applications seems impossible to assess but surely is in the range of six-digit or beyond (for example the website Wikia.com, as the largest wiki-hosting provider advertises to host more than 200,000 distinct wikis as of 2012 (Wikia, 2012)).

Probably the most fundamental wiki feature is the free and welcome access of any interested user to edit wiki content. The latter is basically represented as **wiki-articles** (hypertext-pages dedicated to the description of a subject), but can also involve most media formats (documents, images, videos, etc.). Editing includes creating, changing, deleting and linking wiki articles in order to collaboratively evolve a knowledge network. The technological barrier is relatively low as users don't need any additional client software but their web browser, the edit mode is directly accessed from an article page and the wiki-markup language (*"wikitext"*) is fairly easy to learn. From there on, functionality between different wiki-engines vary greatly, as the term is applied to a diverse set of systems. The popular MediaWiki (MW, 2012) for example, which is also the wiki-engine used on the Wikipedia, records the complete content history, and any earlier version can be revoked which is as much a security as a documentation feature. Navigation is kept uniform and simple and is supported by additional functions like search, recent changes or the possibility for the user to select specific pages for monitoring. Any article can be assigned to one or more categories, which is used as a simple structuring mechanism. Furthermore, the use of formatting templates is supported which can help to create and maintain an internal (informal) structure of the content.

In knowledge management research wikis are widely perceived as potent knowledge management instruments (e.g. O'Leary, 2008; Wagner, 2006) and practitioners have successfully employed wikis in various environments including organizational and enterprise knowledge management (Grace, 2009; Lykourentzou et al., 2011), software engineering (Clerc et al., 2010; Solis et al., 2009), teaching (Himpsl, 2007; Parker & Chao, 2007), or also in more exotic use cases, as for example in disaster management (Yates & Paquette, 2011). Also in the IWRM domain, some organizations have recently started initiatives of which probably the most visible examples are the UNDP-initiated WaterWiki (WaterWiki, 2012) and the IWAWaterwiki (IWAWaterWiki, 2012).

The academic world has been somewhat reluctant to accept wikis as information sources and open knowledge management tools, primarily due to quality concerns and the anonymity of authorship. On the other hand, Giles (2005) early demonstrated a high quality of scientific articles in the Wikipedia by a blind accuracy comparison to those in Encyclopaedia Britannica. Also with moderated or even peer-reviewed wiki-clones (e.g. *"Scholarpedia"* (Izhikevich, 2006) or the pioneering *"Nupedia"* project (*not existent anymore*)), scientists have proceeded to meet the concerns on "wiki-risks" (Black, 2008). And consequently, Wikipedia and other wikis are increasingly used as academic reference (Park, 2011). As with many knowledge management approaches, out of the academic spectrum it was especially in the field of life sciences that wiki technology was early adopted within specialized communities (Hoffmann, 2008).

Surely, there have also been several unsuccessful wiki-applications which, however, are not very well documented in the literature (for one example see (Cole, 2009)). On the other hand, most of the mentioned authors identify knowledge management strength as well as weaknesses in wikis (**Table 2.4**). Most perceived limitations of conventional wikis are rooted in a lack of formalized structure (limitations in bold font in Table 2.4).

From a (knowledge-) structural viewpoint a wiki-page is a subjective and informal collection of information elements residing implicitly in the linguistic content. Also the hyperlinks between pages do not add any semantic information because every link has the same meaning. For example, a link that refers from the wiki-article on the city of As-Salt to the article on Jordan has the exact same explicit meaning as the link pointing back to the wiki welcome page (**Fig. 2.9**a). It is thus not possible for a machine to recognize the meaning of that link as "located in" and to answer a question for all cities located in Jordan. Answering such queries typically is done by using manually crafted and updated lists.

Table 2.4: Strengths and limitations of wikis according to authors cited in this section and seen within the classic knowledge processes after Holzner and Marx (1979).

Process	Strengths	Limitations
Knowledge Acquisition	- Collaborative generation - Unlimited Contributors - Constant review and updating - Low technological barrier to contribute	- Lack of content quality control - **User not supported in structuring** - Easy to introduce a bias - **Information import not supported**
Knowledge Storage	- History creates full documentation - Evolutionary user-consensus - User controlled content and organization	- **Knowledge representation is mostly implicit in the content** - **Information between articles can be ambiguous** - **Keeping content updated is work intensive** - **Information can easily get lost in large wikis (abandoned pages)**
Knowledge Dissemination	- *One-to-many* and *many-to-many* communication - Web-based - Very low technological barrier to access	- **No queries - Information must be extracted manually** - **No alternative views on information possible** - **Mostly only fulltext-search**
Knowledge Application	- Linking allows to reuse and evolve knowledge within wiki	- **No machine-readable structure to reuse outside the wiki**

Consideration of these limitations has led to semantic wiki development initiatives. Semantic wikis basically expand the wiki-technology with the possibility to create a formal semantic knowledge representation of the wiki content by annotating the content on the wiki-pages with a special markup. An overview of general architectural features can be found in Oren et al. (2006).

The level of formalization of page-content that is realized with such annotations varies between the different wiki-engines as well as with the purpose of the wiki. The typical approach is to represent one ontological concept (e.g. a city, a spring, etc.) as one wiki-article. Formalized statements about the concept are made by annotating article content as "typed links", which include information about the character of the relation.

Using the example from above, the hypertext link between the As-Salt and the Jordan articles would be annotated as in **Fig. 2.9**b, thus formally stating that As-Salt has an attribute (*located in*) which can be given a value (*Jordan*). Hereby, most semantic wikis internally represent the graph of typed pages and links in RDF/OWL (or some subset) (Buffa et al., 2008). This allows differentiating between typed links in regard to the datatype of their object into object properties (object of the typed link is a wiki-page or some external document) and datatype properties (object is a data value).

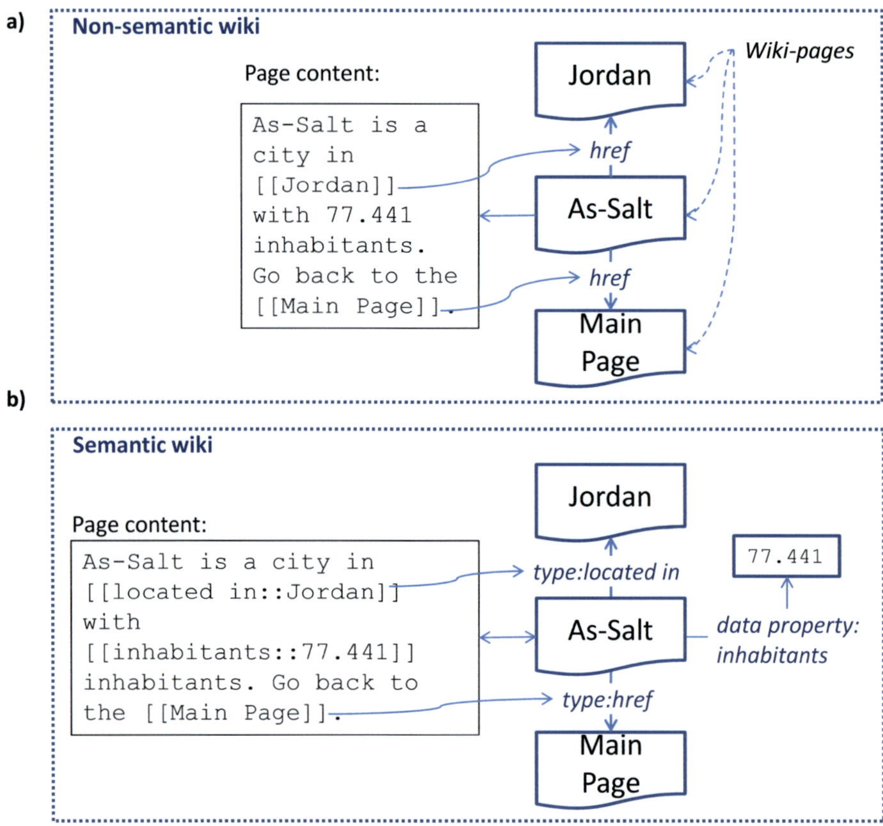

Fig. 2.9: Difference between **a)** ordinary links in non-semantic wikis and **b)** typed links as structural elements in semantic wikis (here using the Semantic MediaWiki markup (SMW, 2012).

Eventually, the sum of statements within the wiki-content creates a machine interpretable ontology ("wiki-ontology"). Using this "wiki ontology" enables a series of supportive features for contributors, such as help to find the right place for their contribution and suggest structural elements (e.g. links to existing content) during contribution. By allowing "inline queries" on the formalized structure it supports better and direct search, various representation possibilities (e.g. automatic generated and updated lists, tables, maps, etc.) and extraction of content. Another advantage of querying is that an information element (e.g. the number of inhabitants) can be kept on one wiki page (e.g. the article of As-Salt) but displayed on other pages (e.g. in a list of Jordanian cities) without copying the number, which helps to keep information consistent.

Several semantic wikis also allow the tagging of pages to express a class membership. **Fig. 2.10** illustrates the mentioned basic building blocks and their interrelations.

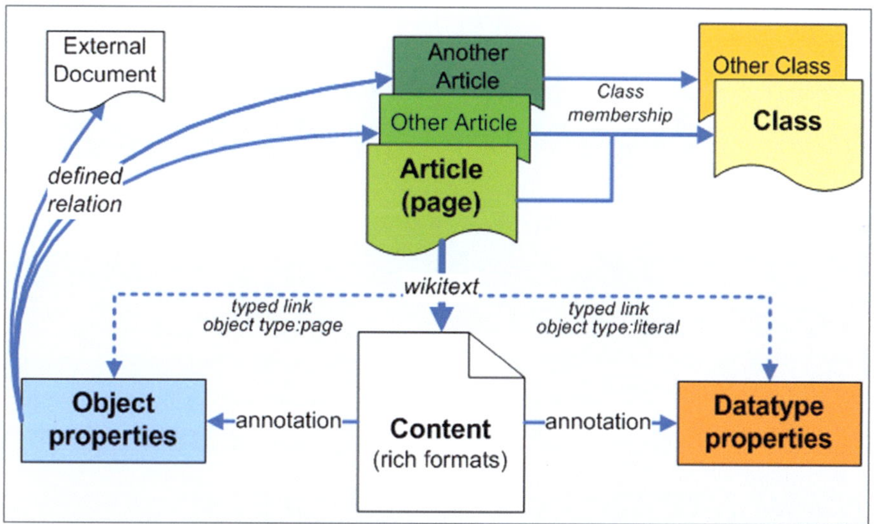

Fig. 2.10: Typical building blocks of a semantic wiki knowledge structure. The wikitext can be annotated to define typed links that formalize content as either data values (datatype properties) or defined relations (object properties) of the subject represented in the article. Tagging of articles also allows defining class membership.

From a knowledge representation perspective, the foremost limitation of some of the existing semantic wikis (e.g. SMW: (Krötzsch et al., 2006)) is the reduced expressive knowledge representation possibility, due to the page-centrism (cf. Chap. 4.3.3 and 4.3.7). From user perspective, the need to learn additional syntax may be a small barrier, although the required functional markup is usually rather limited. A more important barrier in this context probably arises from the necessity for the users to develop a good understanding of the wiki ontology in order to fully use the semantic features.

A good overview on different semantic wiki development projects can be found in Buffa et al. (2008). Examples for semantic wiki engines that have continuously grown and stabilized are the Semantic Mediawiki-project (Krötzsch et al., 2006) and the KiWi-project (Schaffert et al., 2009). In recent years semantic wikis have also found into commercial products as for example in *SMW+* distributed by the semafora systems GmbH (formerly: Ontoprise GmbH), or at Wikia (Wikia, 2012) as the largest wiki hosting site. Discussions to introduce semantic wiki-features to Wikipedia are already on-going for some time (Völkel et al., 2006) and are currently investigated in the Wikidata-project (Wikidata, 2012). There are a row of publications that propose the use of semantic wikis in interesting knowledge management application scenarios (Darari & Manurung, 2011; Krabina, 2010; Wagner, 2006). Scientific publications on the actual application of semantic wikis are still sporadic but publication intervals are steadily increasing. Lange (2011) presents a thorough study on the application of semantic wikis in a framework for mathematic knowledge management. Also the field of life science has shown early adoption again (Cariaso & Lennon, 2012; Li et al., 2012).

*"The important thing in science is not so much to obtain new facts
as to discover new ways of thinking about them."*
(Sir William Henry Bragg, 1862-1942)

3. IWRM-Modelling and Scenario Planning

3.1. Framework for IWRM-Modelling and Scenario Planning

This chapter presents the conceptual development and the application of a sub-basin-scale approach on Integrated Water Resources Management and scenario planning. The conceptual modelling framework employed therefore is illustrated in **Fig. 3.1**.

The IWRM theory usually promotes the basin as adequate management unit. Granting that basin scale scenarios are useful for setting broad national water management strategy goals, they frequently lack reliability due to the non-validated assumptions on their local implementation. The approach taken here emanates from a detailed water system model of the Jordanian Wadi Shueib sub-catchment, which was established on the current state of knowledge of the responsible water sector institutions.

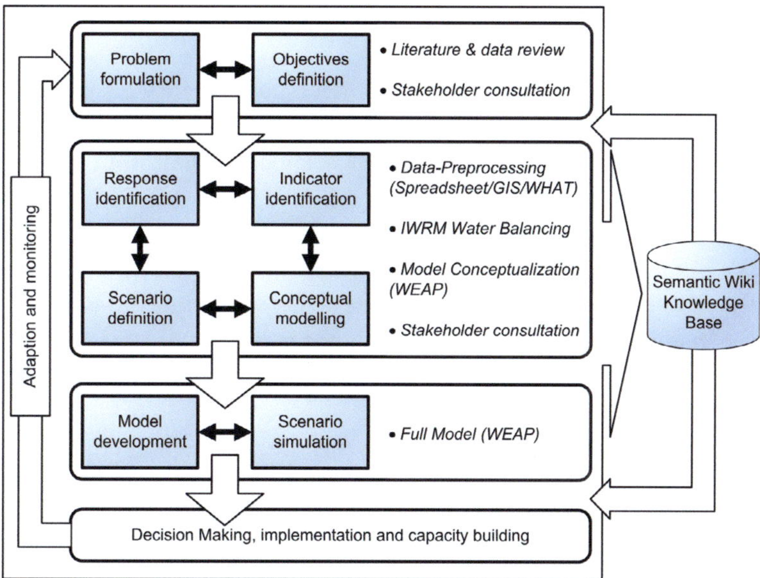

Fig. 3.1: Conceptual framework of the IWRM-modelling and scenario-planning approach developed for the Wadi Shueib use case. Blue marks the steps that were performed in the scope of this thesis.

3.2. Study Area

3.2.1. Selection of the Study Area

The Wadi Shueib catchment in Jordan was chosen as to provide a realistic use case for the development of the IWRM concept and to evaluate the potential of the collaborative and knowledge based approach.

The selection of the Wadi Shueib catchment as suitable case study was for mainly two reasons. First of all, the area faces a set of urgent water management challenges that are set at the intersection of competing municipal, industrial and downstream agricultural demand and at the same time displays concrete water resources pollution problems that are likely to aggravate in the near future. In this regard, the situation in the Wadi Shueib can be considered as a blueprint for many of the problems the region is facing and the challenges to be addressed by the Jordanian Water Strategy.

The second main reason being that, with regard to Jordanian standards, the Shueib area can be considered as one of the more intensely studied sites in Jordan and thus provided access to a range of precursory works related or relevant to water resources topics. Relevant and available studies in the Wadi Shueib area examine (in chronological order):

a)   the discharge and hydrochemistry of the numerous Shueib springs (Ta'any, 1992),

b)   an hydrochemical analysis of aquifer susceptibility to pollution (Abu-Jaber et al., 1997),

c)   the overall hydrological and hydrogeological situation in the catchment (Becker, 2000),

d)   soil characteristics and soil map (Kunz, 2003),

e)   the mapping of water and groundwater hazards (Storz, 2004),

f)   the distribution and concentration of heavy metals in soils samples in the vicinity of the city of Fuheis (Banat et al., 2005),

g)   land use characteristics, groundwater risk and vulnerability maps (Werz, 2006),

h)   isotopes for groundwater recharge estimation (Zagana et al., 2007),

i)   agricultural irrigation water use efficiency (Jiries et al., 2010),

j)   the spring recharge zones and the proposed area for the delineation of spring protection zones (BGR & MWI, 2010)

3.2.2. Geographic Overview

The Wadi Shueib (also: Shu'aib, Shoeib, etc.) is a Jordanian sub-catchment of the Lower Jordan River Basin in the Balqa governorate west of the capital Amman. The catchment defined in this study features an area of 198 km² upstream of the Wadi Shueib Dam. It is characterized by a steep relief ascending from -200 m bmsl in the southwest part up to above 1250 m amsl. in the northeast (**Fig. 3.2**).

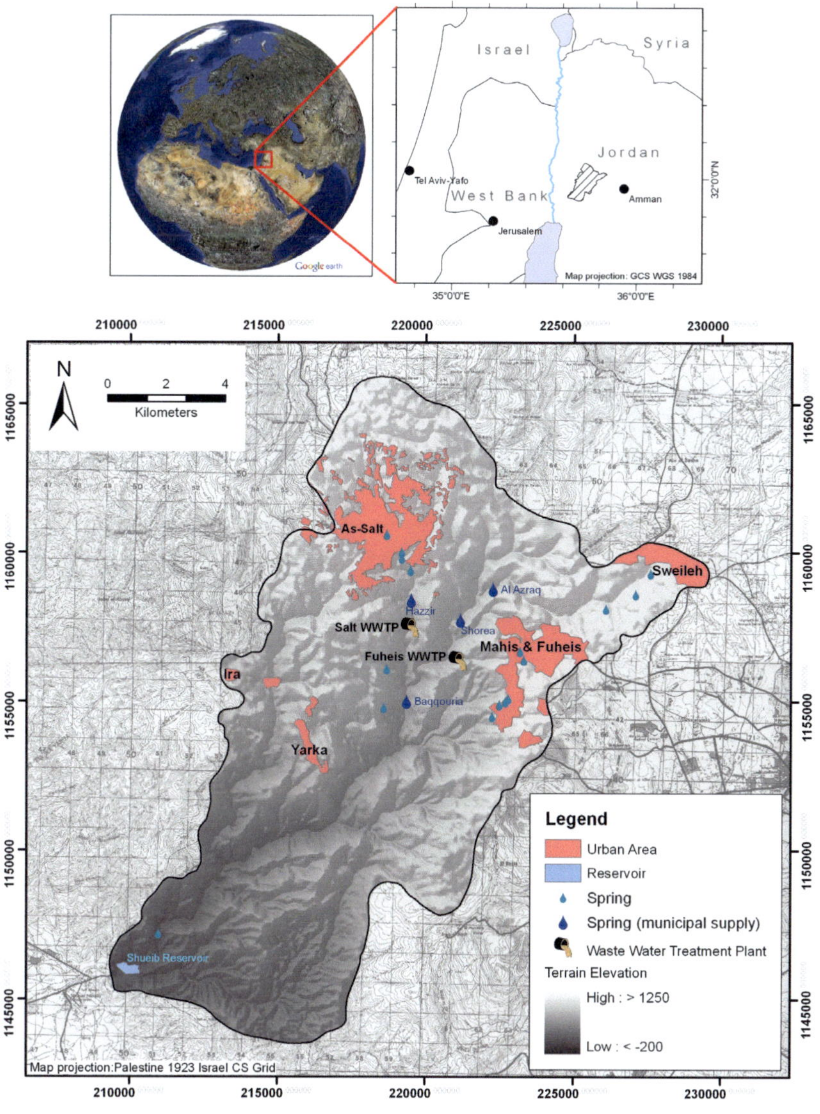

Fig. 3.2: Location and Overview of the Wadi Shueib study area.

3.2.3. Climate

Due to the strong relief, a the climatic conditions vary considerable in Wadi Shueib from the arid Jordan Valley and along the ascent towards a Mediterranean climate in the highlands to the northeast (orographic effect). The major large-scale climatic influence in the region is the Mediterranean circulation with summer months usually without any precipitation and winter months where the dominating south-west winds bring low pressure areas and rainfall from the eastern Mediterranean.

The major amount of rainfall occurs in the winter month from October to March in the higher altitudes to the northeast with annual precipitation amounts often exceeding 600 mm (in 44 % of recorded years) and is decreasing considerably towards the Jordan Valley where annual amounts are usually below 200 mm (in 67 % of recorded years). This tendency is shown in **Fig. 3.3** as spatial interpolation of long term annual precipitation averages calculated from available meteorological station data.

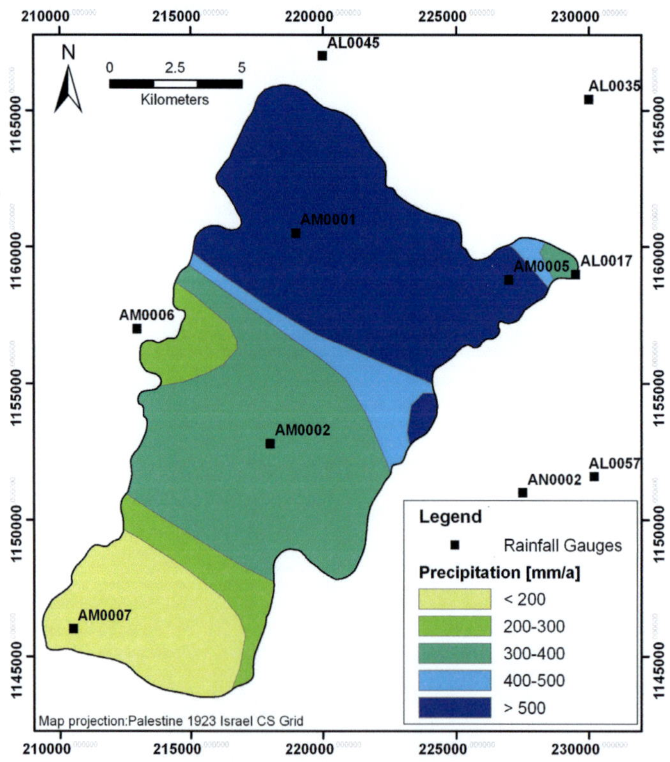

Fig. 3.3: Inverse Distance Weighted interpolation of long term average annual precipitation in the Wadi Shueib catchment area (based on data records from the MWI).

The average temperature shows a contrary trend with daily means rising about 5-10 °C towards the Jordan Valley, where the temperature ranges from around 12 °C average daily minimum in January up to above 40 °C average daily maximum in July.

The profiles in **Fig. 3.4** shows the long term averages of precipitation and temperature for two stations in the high and low altitudes of Wadi Shueib that display the mentioned trends. For the As-Salt station, suitable temperature data was only available for the years 2004 and 2005, which, however, have been climatically average years and are thus displayed here as typical values.

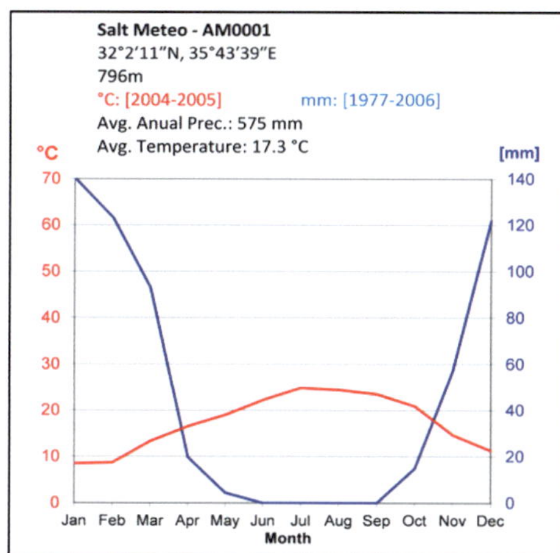

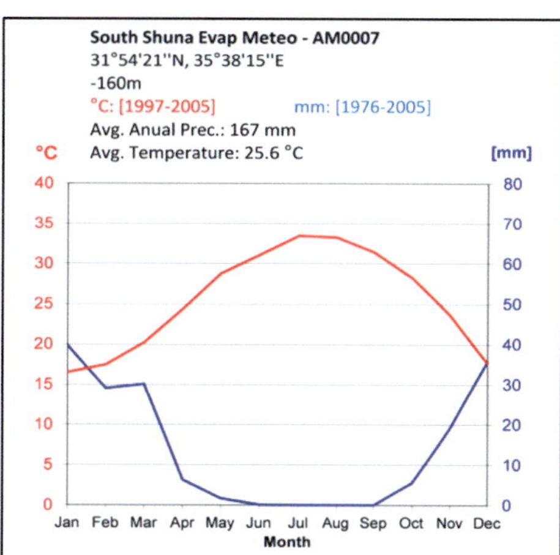

Fig. 3.4: Profiles of average temperature and precipitation at high and low altitudes in the Wadi Shueib catchment. Please refer to **Fig. 2.10** for the exact locations of the meteorological stations (based on data records from the MWI).

In addition to the spatial and seasonal rainfall variability, the region also experiences strong interannual precipitation variability with a standard deviation in annual totals of about 35 % for the Wadi Shueib catchment area (**Fig. 3.5**), which lies somewhat above the regional variability for the Lower Jordan Valley of about 30 % as given by (Black, 2010).

Potential evaporation rates are monitored as Class-A Pan evaporation at some stations in the catchment (AM0007) or its vicinity (AL0035, AL0057, Karama) and also display the differences in relation to the altitudes with higher evaporation rates in the semi-arid Jordan Valley area (**Fig. 3.6**).

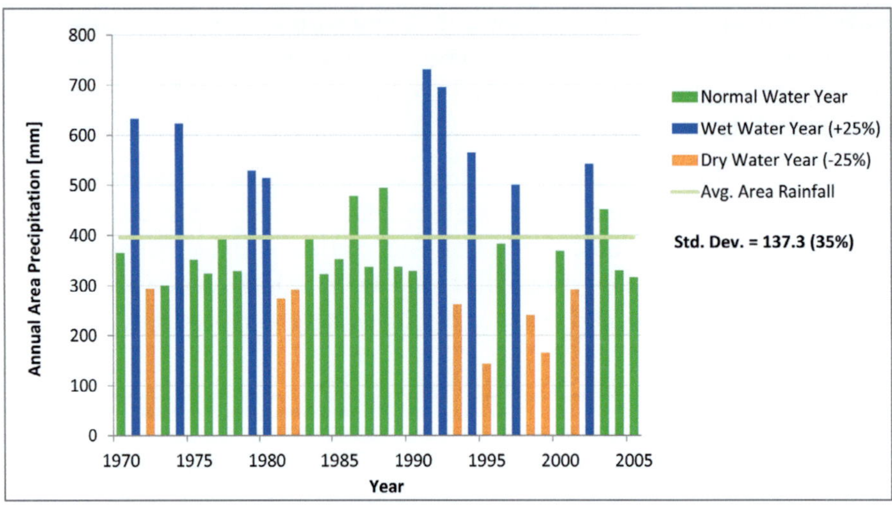

Fig. 3.5: Long term interannual variability of average areal rainfall in the Wadi Shueib catchment. Distinction between wet and dry water years is made arbitrarily by a +/- 25 % threshold.

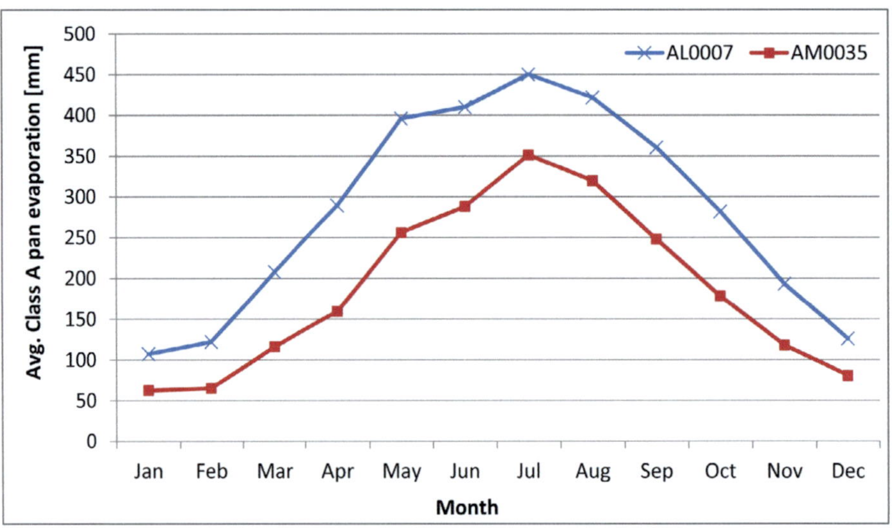

Fig. 3.6: Monthly averages of Class A-Pan evaporation rates in the higher (AM0035) and lower altitudes (AM0007) of Wadi Shueib (based on data records from MWI).

3.2.4. Geology

Geology, stratigraphy and tectonic setting of the Wadi Shueib catchment area have already been described in detail in the works of Werz (2006), Sahawneh (2011) and Hahne et al. (2008). Thus, with reference thereto and the references cited therein, a short overview will suffice at this point.

The following geological maps (1:50,000) and reports from the National Resources Authority of Jordan cover the study area:

- As Salt – 3154-III (Kahlil, 1993)
- Karama – 3153-IV (Shawabkeh, 2004)
- Suwaylih – 3154-II (Barjous, 1993)
- Amman – 3153-II (Diabat, 2004)

Various different taxonomies and classifications have been in use for the description of the areas stratigraphic units (e.g. (Andrews, 1992; Bender, 1974; MacDonald et al., 1965; Masri, 1963; Parker, 1970; Quennell, 1951). The following sections will employ the nomenclature of the Jordan 1:50,000 National Geological Mapping Project (1993-2004) that basically follows (Masri, 1963) and (MacDonald et al., 1965).

3.2.4.1. *Geological Setting*

As side valley on the eastern rift shoulder, the Wadi Shueib is highly affected by the vertical and horizontal tectonic movements related to the formation of the Jordan Rift. The Rift Valley formed as a lateral strike-slip fault system with an estimated 105 km offset (Freund et al., 1970) since the upper Miocene with an initiation approx. 12 MA ago (Bayer et al., 1988). The Wadi Shueib Structure, as the main structural element in the study area, developed as right-lateral shear zone offspring of the Dead Sea Transform Fault (Mikbel & Zacher, 1981). It forms a reverse flexure of approximately 25 km length and 1-4 km width which extends from the Jordan Valley in the south towards northeast and finally bends into a monocline known as Suweileh Structure (Salameh, 1980).

The flexure along Wadi Shueib is accompanied by numerous anticlines and synclines with a general trend of their axes of NE-SW (**Fig. 3.7**).

Normal faults in the Wadi Shueib area display prevailing striking of NW-SE with downthrow to the E in the northern part and to the W in the southern (Becker, 2000) and reverse faults in NNE-SSW striking with a NW vergence (Fig. 3.7). Along the main tectonic faults the vertical displacement is found to be more than 400 m in some places (Hahne et al., 2008).

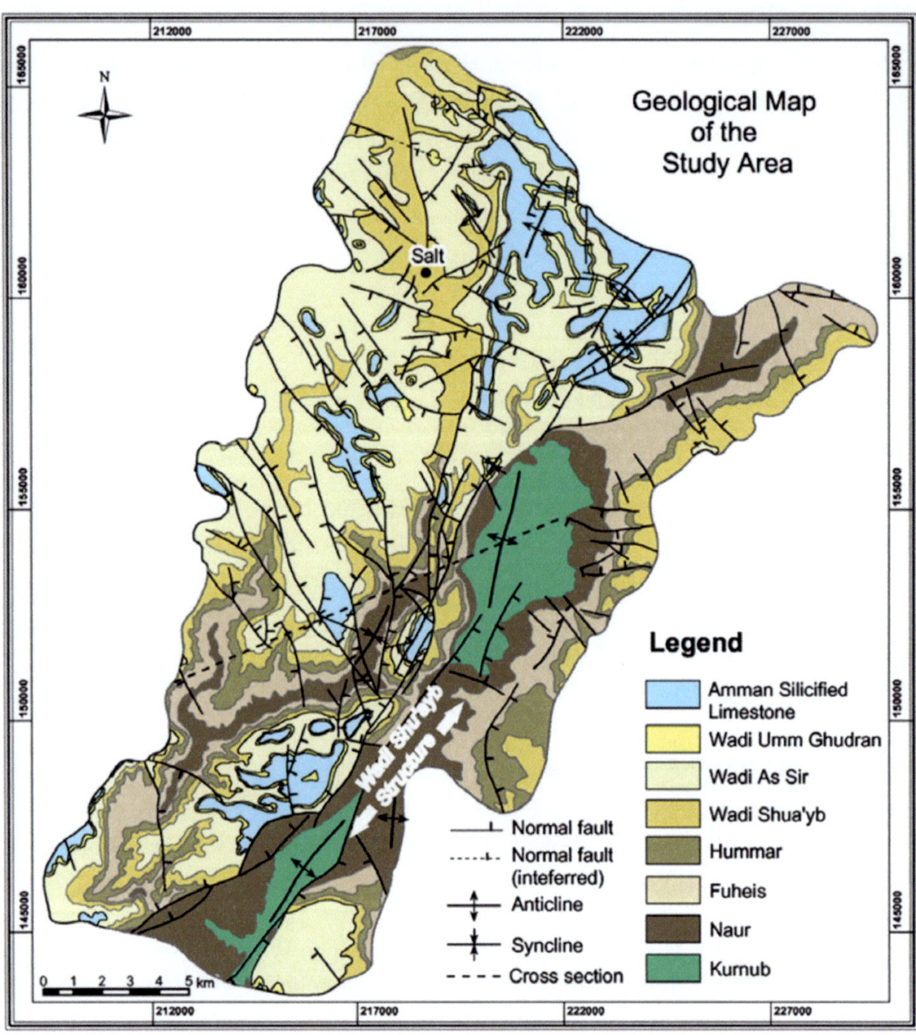

Fig. 3.7: Geological Map of the Wadi Shueib catchment area (Werz, 2006).

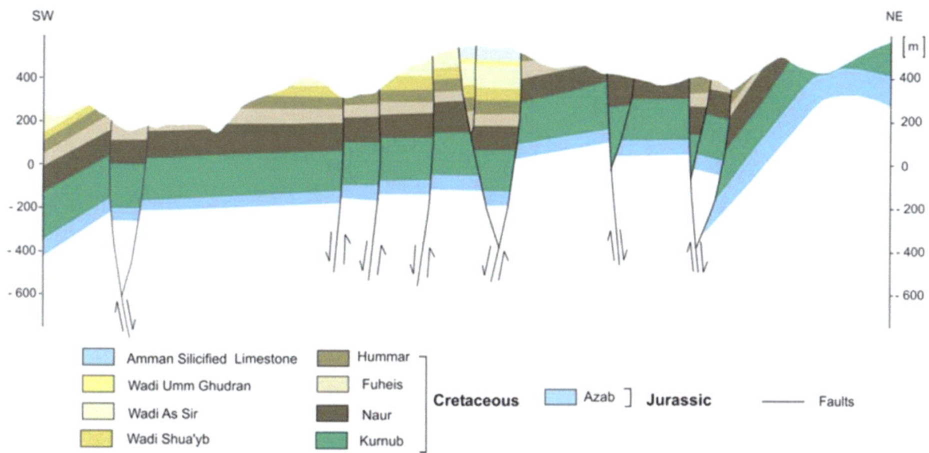

Fig. 3.8: Geologic profile through the study area. The profile location is marked in Figure 3.7 (Werz, 2006).

3.2.4.2.   Stratigraphy

Outcropping rocks in Wadi Shueib consist exclusively of Mesozoic sediments from Early Jurassic (Z2: Azab Group) to Late Cretaceous (Balqa Group) age. The stratigraphic succession in the study area is summarized in **Table 3.1**.

No outcropping of Triassic rocks (Z1: Ramtha Group) is observed in the study area, the next outcrops can be found to the southeast in the Wadi Kafrein area (Lenz, 1999).

Jurassic sediments are found in the area of the Wadi Shueib Structure to the west of Mahis where they expose a thickness of about 40-60 m (Hahne et al., 2008). The Jurassic is represented in Jordan by the Azab Group (Z2) consisting of limestone, marlstone, clay stone and sandstone formations. Triassic and Jurassic together are also termed Zarqa Group (Z1+Z2).

The Jurassic is unconformably overlain by the sandstones of the Lower Cretaceous Kurnub Group (K1+K2) which display extensive outcrops along the axis of the Wadi Shueib structure.

The governing outcropping formations in the Wadi Shueib area are the calcareous sediments of the Late Cretaceous Ajlun (A1-A7) and Balqa (B1+B2a/b *also: Belqa*) Groups, consisting of various forms of limestone, dolomitic limestone, dolomite and marlstone.

Table 3.1: Stratigraphy of the Wadi Shueib. The taxonomy follows the 1:50,000 Geological Maps from the National Resources Authority.and the thickness range is given as found on the corresponding map sheets. The lithology follows the description of Werz (2006). The jagged lines express unconformable sequence overlays.

Epoch	Age	Group		Formation		Lithology in Wadi Shueib	Thickness
Late Cretaceous	Campanian	Balqa (or: Belqa)		B2b	Al Hisa Phosphorite	Phosphorite beds, with layers of chert, limestone and silicified limestone.	10-15 m
				B2a	Amman Silicified Limestone	Silicified limestone and massive chert banks.	40-60 m
	Santonian			B1	Wadi Um Ghudran	Chalk and marlstone, (some marly limestone).	40 m
	Coniacian	Ajlun		A7	Wadi Es Sir Limestone	Massive limestone banks and dolomitic limestone, (chert), (marly limestone).	80-150 m
	Turonian						
				A5/6	Shueib	Marly limestone, chalky limestone, nodular limestone, marls.	80 m
	Cenomanian			A4	Hummar	Limestone, dolomitic limestone, dolomite, (marly interbeds).	40-65 m
				A3	Fuheis	Marl, calcareous siltstone, nodular limestone, marly limestone.	50-80 m
				A2	Na'ur	Limestone and marl overlain by marly limestone. Dolomitic nodular limestones towards the top.	90-200 m
				A1		Limestone and dolomite with interbedded marly, sandy facies towards the base.	
Early Cretaceous	Albian : : Berriasian	Kurnub		K2	Subeihi	Massive, cross-bedded varicoloured sandstone, siltstone layers towards the top.	150-350 m
				K1	Aardo	Cross-bedded medium to coarse-grained white, pale, yellow and pink sandstone.	
Jurassic	Oxfordian : Hettangian	Zarqa Group	Azab	Z2	[Mughanniya], [Hammam], [Ramla], [Dhahab], [Silal], [Nimr], [Hihi]	Sandstone, limestone, dolomite, marl and clayey siltstone.	50-400 m
Triassic	Norian : Olenekian		Ramtha	Z1	[Umm Tina], [Iraq el Amir], [Mukheiris], [Hisban], [Suwayma]	Shales, siltstones, sandstones, marl, limestone, dolomite, oolithic limestone and evaporites. No outcrops in Wadi Shueib area	250-500 m

3.2.5. Hydrogeology

On a regional scale the Jordanian hydrogeology distuinguishes three main aquifer complexes (Salameh & Udluft, 1985):

1) the **Deep Sandstone Aquifer Complex** (Ram – Zarqa – Kurnub System) forms one unit in southern Jordan that gets separated towwards the north by thickening limestone and marl strata.

2) the **Upper Cretaceaous Aquifer Complex** (Ajlun – Balqa Aquifer System) or Middle Aquifer Complex of alternating aquifer/aquitard sequences of limestone, dolomites, marls and chert beds.

3) the **Shallow Aquifer Complex** (Tertiary – Quaternary Aquifer Systems) comprising the limestone, chalk and chert units of the Umm Rijam and Wadi Shallallah formations, the Basalt Aquifer and the Alluvial Deposits of Tertiary and Quaternary Age.

The relevant aquifers formed by the Mesozoic sediments in the Wadi Shueib study area belong to the Kurnub Group of the Deep Sandstone Aquifer Complex (also: Lower Cretaceaous Aquifer) and the Upper Cretaceaous Aquifer of the Ajlun and Balqa Groups. The underlying units of the Zarqa Group are considered as the base of the hydraulic system and are usually regarded as highly fractured aquitard with low primary but high secondary permeability and mostly brackish water content (JICA, 1995; Margane & BGR, 2002). However, until now little is known about the extent, hydraulic conditions and mineralization of this aquifer in the study area.

The Deep Sandstone Aquifer Complex is represented by the outcrops of the Kurnub Group (K1+K2) sandstones along the Wadi Shueib structure. There are no noteworthy springs that emerge from this aquifer in the study area and although it is of significant thickness and usually contains good quality groundwater in the highland areas, it is not exploited for water supply in the study area. In this regard, Margane et al. in (BGR & MWI, 2010) suggested that an exploitation of the aquifer may "...considerably reduce the risk of bacteriological contaminations [in the water supply] because it is covered by aquitards and therefore naturally protected against contamination...when major fault zones are avoided."

The Upper Cretaceaous Aqufer Complex is represented by the interstratifications of limestones, marls and dolomites of the Ajloun and Balqa Groups (A1-A7 and B1-B2). Within this sequence the three aquifer subsystems of the Naur Aquifer (A1/2), the Hummar Aquifer (A4), and the Wadi Es-Sir Aquifer (A7) are interbedded between the Fuhays (A3) and the Shueib (A5/6) Formations, which act as leaking aquitards that are hydraulically connected through fractures and fault zones (Werz, 2006).

The limestones and dolomites of the Upper Cretaceaous Complex exibit different levels of karstification as descibed by (Werz, 2006): The Naur limestones (A1/2) are moderately karstified with some smaller caves and small karren. The massive cliff strata of the Hummar (A4) formation favour a strong karstification where angular cave formation and dissolution features along joints and bedding planes are common. The resultant high permeability makes A4 an important aquifer in the Wadi Shueib despite the relatively small recharge area. The Wadi Es-Sir Limestone (A7) shows high to very high degree of karstification of the micritic limestone with abundant caves, and distinct surface karst features such as karren.

The A7 formation receives the major part of the rainfall recharge, as it is the top layer and exhibits the most extensive outcrops especially in the higher altitudes where the average annual precipitation is high. The aquifers below consequently receive limited (A4) or almost no (A1/2) direct recharge from precipitation.

There is no available network of wells for groundwater monitoring in the area and consequently no groundwater contour map is available. Margane et al. in (BGR & MWI, 2010) assumed that the hydraulic system is well interconnected and that water levels in the different hydraulic units are more or less identical at a location controlled by topography and degree of fracturing.

Only little has been published on the hydraulic properties of the Wadi Shueib aquifers and no detailed studies have been conducted hereto. However, a number of reports provide estimations for the general hydrogeological characteristics of the formations. As an overview, **Table 3.2** summarizes the geologic succesion and given general estimations on the aquifer potential of the relevant aquifer systems in the Wadi Shueib area.

Table 3.2: Hydrogeological classification in Wadi Shueib. The hydraulic properties of the hydrogeological units are general estimates documented in the following reports: [a] Margane and BGR (2002); [b] Geyh et al. (1985); [c] Parker (1970); [d] Salameh & Udluft (1985); [e] JICA (1995); [f] Abu-Ajamieh (1998); [g] (Al-Kuisi, 1998).

Complex	Group	Formation	Class	Hydraulic Conductivity	Storage Coeff.	Transmissivity	Comments
Upper Cretaceous Aquifer	Belqa	A7-B2	Aquifer	$2*10^{-5}$ m/s [a] $7*10^{-4}$ m/s [b] 10^{-3} - 10^{-4} m/s [c]	0.01-0,1 [c] 0.001 [f]	0.3 - 46000 m²/d [c]	A7-B2 are regarded as one karstic aquifer system. Extensive outcrops in western highlands with probably high recharge. Most productive aquifer in Wadi Shueib.
	Ajlun	A5/A6	Aquitard	10^{-9} m/s [a]			
		A4	Aquifer	$2*10^{-5}$ m/s [a] $4*10^{-4}$ m/s [d] $7*10^{-4}$ - 10^{-5} m/s [c, f]	~0.005	230 - 2800 m²/d [c]	Karstified limestone with good aquifer potential in Wadi Shueib, but lowest sum of discharge of the three aquifer systems
		A3	Aquitard	10^{-9} m/s [a]			
		A1/A2	Aquifer	10^{-5} m/s [a] $3*10^{-5}$ [b]		3 - 10 m²/d [c] 142.8 m²/d [e]	More springs are discharging from this aquifer in comparison to A4, in other areas it is considered a poor aquifer but it feeds the largest spring in the Wadi Shueib area.
			Aquitard	$2*10^{-8}$ m/s [b]			
Deep Sandstone Aquifer	Kurnub	K1/K2	Aquifer	$2*10^{-5}$ m/s [a, d, e, f] $5*10^{-4}$ - $7*10^{-6}$ m/s [c] $5*10^{-4}$ - $5*10^{-5}$ m/s [g]	~0.002 [e] ~0.0001 [f]	110 - 120 m² /d [e] 23.2 - 84 m²/d [f] 27 - 260 m²/d [c]	Sandstone Aquifer. No significant springs emerge from the Lower Aquifer Complex in the Wadi Shueib area
	Zarqa	Z1/Z2	Aquitard / Aquifer	$1.5*10^{-5}$ - $3*10^{-6}$ m/s [e]	~0.001 [e]	60 - 160 m²/d [e]	Artesian aquifer of sandstone, siltstone, limestone and shale. Low primary, but high secondary permeability. Hydraulically connected with Kurnub Sandstone.

3.2.5.1. **Wells**

A number of wells were drilled in the Wadi Shueib, especially in the area around As-Salt and to the south near the Shueib Dam (**Fig. 3.9**). All wells are exclusively penetrating the Hummar (A4) and the Naur (A1/2) aquifer. Most are governmental and intended for drinking water production for the local communities and some areas to the north and the south of the study area. Formerly private wells have either been taken under governmental control or sealed. Many of the wells in the area have, however, already been inactive for a long period due to regular occurrences of high nitrate concentrations. **Table 3.3** lists the wells that have recorded abstractions for drinking water production during the last decade. Today, the only remaining active well that is used for direct supply within the study area is the Salt Municipality Well No. 4 (AM1004), with an average productive yield of about 0.33 MCM per year, whereas the annual 1.21 MCM produced from the Yazidiyya Wellfield are primarily allocated for supplying the rural communities in the Zai area to the north. But in high demand times the system allows this water to be pumped to the As-Salt Network. The Jrea'a Wellfield is exclusively supplying downstream networks in the area of South Shuna in the Jordan Valley.

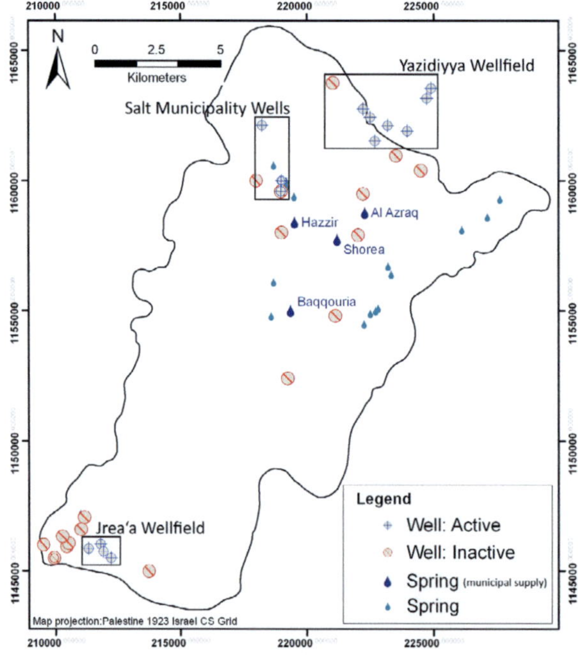

Fig. 3.9: Registered springs and groundwater wells in the Wadi Shueib area and their current status of use.

Table 3.3 List of groundwater-wells in Wadi Shueib that have been used for abstraction in the period between 2000 and 2010 (Source: WAJ-Well Catalogue).

Group	ID	Aquifer	Status
Yazidiyya Wellfield	AL1627	A4	active
	AL2423	A4	active
	AL3525	A1/A2	active
	AL3526	A4	active
	AL3527	A4	active
Salt Municipality Wells	AM1002	A4	until 2002
	AM1004	A4	active
	AM1042	A4	until 2002
Jrea'a Wellfield	AM1020	A2/A1	until 2006
	AM1022	A4	active
	AM1024	A4	active
	AM1026	A2/A1	until 2001
	AM1027	A4	active
Other	AL3459	??	active

3.2.5.2. *Springs*

Springs have traditionally been and often still are the primary water supply in the Jordan Valley side wadis. There are 27 registered springs emerging in the Wadi Shueib catchment area, of which 21 have reported discharges for the period from 2000-2010 (the ones depicted in **Fig. 3.9**), while the rest has apparently run completely dry (**Fig. 3.11**). All of the springs drain the Upper Cretaceous Aquifer Complex and therein the majority in number and discharge volume emerge from the limestone of the A7/B2 and the A1/2-Formation (cf. Table 3.2), while the A4 aquifer appears less relevant (**Fig: 3.10a**). It is assumed, however, that due to the intensive faulting in the area, the A4 aquifer is often in direct hydraulic contact with the underlying aquifer and thus its water is mainly discharged through the springs in the A1/2 aquifer (BGR & MWI, 2010).

The majority of springs in the area show pattern of intermittent discharge with annual peak flows below 100 m³/h occurring in March and April. All of these smaller springs are used by the local land owners, mostly for agricultural purposes. The four most productive springs in Wadi Shueib are the perennial discharging Ain Azraq-Fuheis (A7/B2), Ain Baqqouria (A1/2), Ain Hazzir (A7/B2) and Ain Shorea (A7/B2) with average annual discharges between 1.15 MCM (Hazzir) to 3.42 MCM (Baqqouria). In years with average precipitation amounts, the maximum discharges for these four springs are also observed around March and April about one to two months after the typical precipitation highs in January and February. The discharges of the Azraq and Baqqouria appear closely related to the rainfall and usually show a prompt response to above average rainfall events, while the Hazzir and Shorea springs display a more constant discharge pattern (**Fig: 3.10b**).

Fig. 3.11: Small and large spring in the Wadi Shueib area.

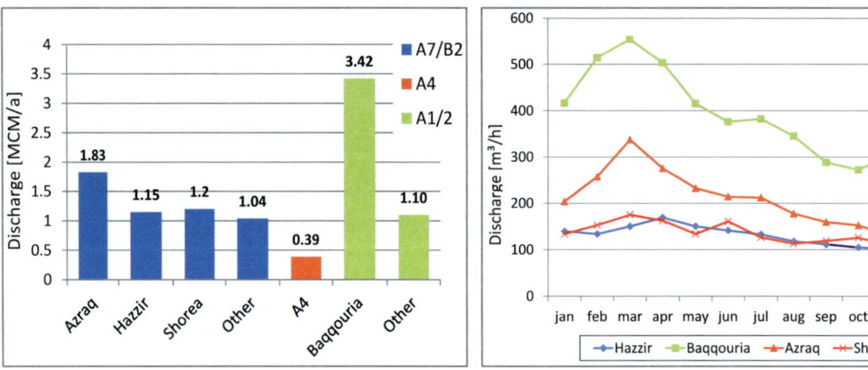

a) Average annual discharge of the Shueib springs in relation to their aquifer of origin.

b) Average monthly discharge pattern of the main springs in Wadi Shueib.

Fig: 3.10. Average annual and monthly discharges of the Wadi Shueib springs over the period from 1980-2010 (based on data records from the MWI).

More details on the spring hydrographs are given by Ta'any (1992) who conducted a comprehensive analysis of the springs in the Wadi Shueib area. According to the data records from the Ministry of Water and Irrigation, the long term discharge of all springs in the catchment is about 9.9 MCM in a normal water year (years with an area precipitation within a 25 % threshold of the long term average).

As expected from the carbonate aquifer origin, the natural water quality of the springs in the Wadi Shueib catchment is classified as earth-alkaline water with high alkaline component and prevailing bicarbonate and some occurrences of sulphate dominance (Becker, 2000; Ta'any, 1992) and salinity ranges between 400 and 900 µS/cm (Abu-Jaber et al., 1997). The limited variation in water chemistry is not unusual with the given lithological similarity between the three aquifers. With these characteristics the spring water provides suitable quality for drinking water supply according to the Jordanian Standards. Thus, the four major springs are used as drinking water supply for the municipalities in Wadi Shueib and some smaller communities in its vicinity. Water from the Azraq-Fuheis spring is directly pumped to the Fuheis Reservoir and used to supply the towns of Mahis and Fuheis, as well as parts of Hummar, Dabouq and Sweileh. There is also a pipeline to the Shorea Reservoir as emergency supply for the Salt directorate. Water from Baqqouria, Hazzir and Shorea is pumped to the Shorea Reservoir at the Shorea Spring and distributed from there towards As-Salt, Ira, Yarka and rural communities in the Zai and Wadi Kirad area.

Different studies found, however, that the aquifers in the Wadi Shueib area are particularly susceptible to pollution from domestic sewage leakage (Abu-Jaber et al., 1997; BGR & MWI, 2010; Ta'any, 1992; Werz, 2006). And Abu-Jaber et al. (1997) pointed out that especially the A1/2 and the A7/B2 render a high vulnerability in this regard as indicated by the significantly higher concentrations of nitrate than found the A4 aquifer, as well as their strong correlation with salinity, Na^+ and Cl^--concentrations.

Since 2001 the water at the Shorea Reservoir is treated in a continuous micro-filtration membrane plant after the spring discharge had become infected with fecal coliform, cryptosporidium and giardia and was halted in 1998 (Saqr, 2001). The Salt Water Treatment Plant can provide a total of 6.5 MCM of potable water per year but often runs on reduced service due to the limited spring discharges, especially in the summer months.

Events of high Nitrate concentrations and microbiological contamination are still prevailing in the spring water of most springs in the area and especially at the Hazzir spring which is regularly exceeding the Jordanian Drinking Water Standard (50 mg/l) in terms of nitrate concentrations (60-80 mg/l in the spring water)(Grimmeisen et al., 2012).

3.2.6. Hydrology

3.2.6.1. Catchment Drainage and Runoff

An often encountered indication of drainage efficiency was introduced by Horton (1932) as the Drainage Density D_d:

$$D_d = \frac{L}{A}$$

where L is the total length of all stream channels (including all channel orders) in the drainage and A is the area of the drainage basin.

The drainage density is usually associated with the rainfall regime and the topography and the lithological infiltration capabilities of the drainage area and gives a qualitative and preliminary indication of the amount of rainfall that is translated into surface runoff (Gregory & Walling, 1976; Schumm, 1977). A high drainage density in this regard indicates a high amount of rainfall-runoff, high flood peaks and high sediment transport, whereas a low drainage density indicates in higher infiltration rates (Singh, 1989).

Drainage density ranges from less than 1 to over 600 km/km² and has been found especially high in semi-arid areas (Gregory & Walling, 1976). Typical values in hard rock terrains are between 1.5 and 6 km/km², in areas composed of carbonate rocks it is usually considerably lower and often below 1 km/km² (Singhal & Gupta, 2010).

For the Wadi Shueib catchment a drainage density of 1.13 km/km² was calculated on the basis of the digital elevation model (DEM) compiled from the 1:20,000 topographic map of the area using the Spatial Analyst module of ArcGIS© 9.3. This is a moderate value in relation to other comparable carbonate drainage areas (Segura et al., 2007). However, the drainage density could be underestimated as the 1:20,000 topographic map might not accurately display all channels of the drainage network.

The majority of surface runoff in the Wadi Shueib drains towards the Dam via the main Wadi channel which has a natural perennial discharge fed by the constant base flow from the larger springs. Today, however, as the major part of the spring water is abstracted for human use in the upper catchment area, the summer base flow especially in dry years consists almost completely of effluent of the waste water treatment plants of As-Salt and Fuheis.

During the winter rains storm the dendritic drainage network of the catchment contributes to the runoff in the main channel. During these flashfloods the wadi runoff often suddenly increases about 20-30 times in comparison to the previous day discharge, according to the inflow records from the Wadi Shueib Reservoir (**Fig. 3.12**).

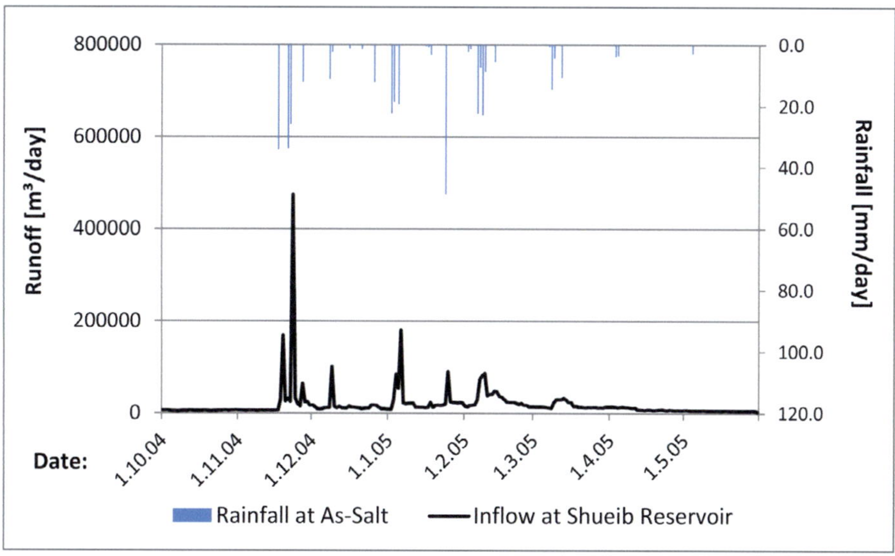

Fig. 3.12: Recorded rainfall at As-Salt station (AM0001) and inflow at the Wadi Shueib Reservoir for the winter 2004/2005 (based on data records from the MWI).

3.2.6.2.   Reservoir

The rainfall flush floods (direct runoff) are stored at the wadi outlet in the Wadi Shueib Reservoir (**Fig. 3.13**) together with the wadi base flow from the springs and the treated effluent from the waste water treatment plants of Salt and Fuheis.

The Dam was originally designed for a storage capacity of 2.4 MCM, but has actually always been operated with 1.43 MCM, due to the seepage losses through the gravels of the dam base layers. The recharged groundwater is pumped from downstream private wells for irrigation purposes in the Jordan Valley shallow aquifer. Therefore, the dam is regarded as an artificial recharge facility of the groundwater as well as a rainfall-runoff surface storage.

As illustrated in **Fig. 3.14**, during the rainless summer months between June and November the reservoir is usually empty and the remaining wadi base flow is left to flow freely through the gate. The Surface outflow of the dam is allocated for irrigation purposes through an open channel carrier canal that serves about 250 ha immediately downstream of the dam. The water is free of charge and each farm owns a share of this water due to traditional water rights.

Fig. 3.13: Wadi Shueib Dam and Reservoir.

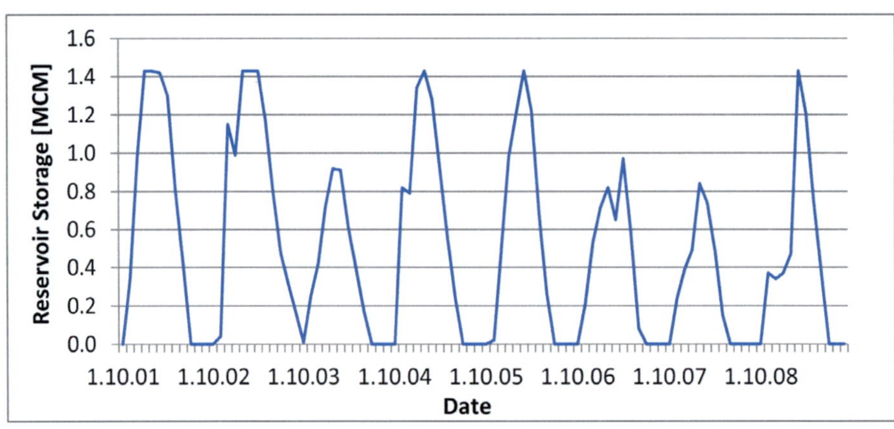

Fig. 3.14: Observed monthly runoff storage in the Wadi Shueib Dam between 2001 and 2008 (based on data records from the MWI).

3.2.7. Population and Land Use

Due to the topography and the climatic conditions, the population density, as well as most agricultural and industrial activity is concentrated in the higher altitudes in the north-eastern part of Wadi Shueib.

The area comprises 5 larger municipalities (Salt, Fuheis, Mahis, Yarka, Ira) and several smaller hamlets. Wadi Shueib has shown a strong demographic development during the last decades. The total population at the last census amounted to 108385 inhabitants and has shown an average yearly population growth of 2.97 % since the last census in 1994, which has been above the average of the Balqa Governorate (2.29 %) and the national average of Jordan (2.60 %) as represented in **Table 3.4** (DoS, 1994, 2004). The noticeably low demographic growth in the city of Fuheis appears related to the consistently reported pollution problems evoked by the local cement quarry and factory.

Table 3.4: Demographic Development in Wadi Shueib between 1994 and 2004

	As-Salt (Sub-District)	Mahis	Fuheis	Ira & Yarqa	Wadi Shueib total	Jordan
1994	56,458	8,000	10,098	6,319	**80,875**	**4,139,458**
2004	77,441	10,649	11,641	8,654	**108,385**	**5,350,000**
yearly growth	3.21 %	2.90 %	1.43 %	3.19 %	**2.97 %**	**2.60 %**

The largest industrial activity in the area is the cement quarry and factory of the Lafarge Jordan Cement group near the city of Fuheis, established in 1951. The plant has a yearly production capacity of about 2.9 million tons of cement. Due to the global crisis in construction and real estates, the production in 2010 was running on less than half capacity (Lafarge Cement Jordan, 2010). The plant has also been a constant political issue for at least the past decade, as especially the inhabitants of Fuheis claim to experience numerous health problems by emitting clouds of dust and chemicals from the factory (Ben Hussein, 2010). A study by Banat et al. (2005) recorded elevated concentrations of heavy metals in soils in the vicinity of the cement factory. The plant operates its own groundwater wells and also uses water from the Tureim spring. According to the annual reports of the company the water used in production amounts to 340 liters per ton of cement.

Other industrial activities in the area are primarily situated within the municipalities. A mapping of these activities can be found in Storz (2004).

The agricultural activity in the Wadi Shueib is mostly situated in the northern highland part of the catchment area and comprises primarily rainfed olives orchards, pasture grounds and fruit gardens. Irrigation is applied where water is available from the smaller springs in the area or along the main wadi stream. About 0.2 MCM per year of treated waste water are directly allocated towards small farming plots in the vicinity of the treatment plants (WAJ, personal communication). The absolute area of irrigated land is difficult to assess and there are no official records as the irrigation is mostly self-organized and self-supplied. A manual classification from satellite images was used to estimate the area of denser vegetation along the wadi channel, known to be used for irrigated agriculture, to about 4.5 km².

3.2.8. Water Supply and Sanitation

Primary water demand in the study area is for municipal drinking water. It is supplied through spring water, groundwater wells and imports and distributed via central networks in the municipalities. The water supply is intermittent and the municipalities receive pumped water from the water authority for (reportedly) 1 to 3 days per week, which is then stored in tanks, either on the roof of the property or underground.

The per capita consumptive use of municipal water for the Balqa Governorate is estimated from the volume billed by the Water Authority of Jordan and accounted to an average of 86 litres per capita per day for domestic uses and 24 litres per capita per day for non-domestic municipal uses (e.g. public buildings and areas) (MWI, 2004). However, the water supply infrastructure is prone to high rates of losses. The average Unaccounted for Water (UFW) in the period from 1997 to 2004 in the Balqa Governorate was estimated by the Water Authority of Jordan as 57.2 % of annual network supply, whereof equal shares are assumed to physical and administrative losses, the latter including non-revenue public uses, illegal connections and wrong metering (MWI, 2004). This considerably contributes to the supply deficit in the area and results in the necessity for households to buy additional water from the water tankers of private vendors which is about ten times the price of tap water.

The water and waste water fee in the Balqa Governorate is at a flat rate tariff of 4.42 JD/month for low volume users (<20 m³ per month) corresponding to 0.221 JD/m³. The fee rises block wise with consumption with a top rate of 160 JD/month (1.24 JD/m3) for users consuming more than 130 m³ monthly.

Part of the municipal waste water is centrally treated at the waste water treatment plants of As-Salt and Fuheis. Especially the smaller villages (e.g. Ira and Yarka and the rural areas) are not at all connected to the central sewer system and households typically employ cesspools and septic tanks. And even within the cities some people still use septic tanks to avoid paying the cost of a sewage connection (Trappe, 2007). The sewer connection ratio for the As-Salt sub-district was estimated at 66 % and at 72 % for the Fuheis-Mahis sub-district (WAJ, 2006). In recent years the WAJ has undertaken considerable efforts to increase the sewer connectivity, especially in the area of As-Salt (MWI, 2010). The plants discharged about 2.1 MCM/a (in 2003) of treated effluent to the wadi, where it is mixed with the base flow and used downstream for unrestricted irrigation from the Wadi Shueib Dam. When the stream reaches the Shueib Dam, over 10 km away, the treated effluent might at times constitute more than 50 % of the dry weather flow.

Households waste water tanks require regular emptying by suction trucks for which house owners have to pay. A household survey on sanitation levels and water awareness in the Balqa Governorate undertaken by the GTZ (in cooperation with the MWI) in 2007 found that 58 % of cesspit owners in As-Salt never use this service and only 23 % do so regularly (GTZ in *Borgstedt and Subah (2008)*; **Fig. 3.15**). The local solution seems to crack the basis of the cesspits and tanks so that the domestic waste water leaks directly through the karstified rock into the underground. It appears sensible to conclude a direct causality to the quality issues reported at nearby freshwater springs, even if no detailed tracing has been conducted yet.

Table 3.5: Basic operation data from the Wadi Shueib waste water treatment plants (MWI, 2004).

	Type of treatment	Capacity m³/d	Population in sewered zone	Population served	BOD Removal	Cost fils/m³ (2003)
As-Salt	EA+MP	7,700	75,000	49,950	94 %	107
Fuheis	EA+MP	2,400	25,000	17,350	94 %	164.5

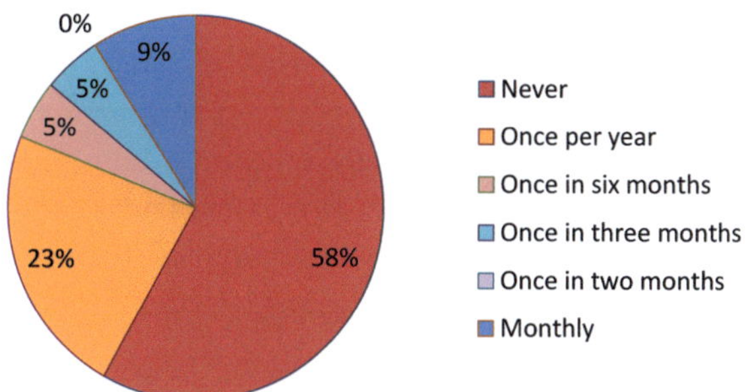

Fig. 3.15: Frequency of waste water tank emptying in As-Salt according to a household survey on sanitation levels and water awareness (data based on and illustration (modified) after GIZ *in Borgstedt and Subah (2008)*).

3.2.9. Water Sector Challenges in the Wadi Shueib

The primary water sector challenges in the Wadi Shueib catchment are related to municipal supply as largest demand sector, as well as to the management of the resulting waste water return flows.

The springs in the area, as the traditional source of local drinking water supply show continuous pollution problems for an already long period that are most likely related to pollution of the aquifers by municipal waste water (cf. Chap. 3.2.5). A multiplicity of additional water and groundwater hazardous activities have also been observed in the study area ranging from the disposal of slaughterhouse leftovers to careless oil spills (Storz, 2004; Trappe, 2007; Werz, 2006). The constantly high population growth (cf. Chap. 3.2.7) is likely to employ increasing pressure on the system and eventually lead to an aggravation of the water quality problems.

Although the quantity of local freshwater would suffice to cover local demands if the quality was acceptable, the municipal network supply has been intermittent during the last decade, often providing water only once per week. Agricultural demands in the studied area itself are of minor extent as the majority of small plots in the upper part of the catchment are rainfed cultures, but ecological hazards are still perceived due to the application of fertilizers in the vicinity of the important freshwater springs due to the lack of enforced spring protection zones (Storz, 2004; Trappe, 2007). Important factor in terms of quantitative agricultural water demand are the downstream farms in the Jordan Valley that receive their irrigation water from the Shueib Reservoir according to their traditional water rights. Any alteration of the water system in the catchment should consider also the impact on these farming families. Even though the national water strategy deliberately states the goal to reduce agricultural supply (RCW & MWI, 2009), it is still questionable if such decisions can be politically viable.

Only limited information has been available on the actual industrial demand in the Wadi Shueib area, but it appears to be of subordinate importance. Yet, the large quarry and cement factory in Fuheis poses a significant hazard to the environment and the water resources in the catchment and also is a constant source of political conflict between the residents of Fuheis and the Lafarge Cement company which operates the site (Lafarge Cement Jordan, 2010).

3.3. IWRM-Model for the Wadi Shueib

The primary objective of the modeling approach in the Wadi Shueib catchment was to develop a case study to test the IWRM planning possibilities as they are given to the local decision makers on the basis of the currently available data and information pool. Secondly the model should eventually act as blueprint for the development of the planned IWRM knowledge management approach by demonstrating the generic structure and the information necessities of such an interdisciplinary planning process. A third objective of the modeling exercise was the exposure of critical knowledge gaps and the evaluation of the resulting uncertainties. According to these objectives it was deliberately decided to base the modeling approach solely on the current data basis and conducting new measurement campaigns.

3.3.1. Modelling Approach

The conducted modelling approach follows widely accepted standards of good modelling practice (Refsgaard et al., 2005), where generally the understanding of the actual state and the urgent challenges of the observed environmental system is perceived as a crucial starting point. This first stage comprises the formulation and identification of specific focus problems, particular objectives and possible response options through a review of all available information and strong stakeholder participation (cf. Chap. 3.2). The subsequent workflow after this initial phase is illustrated in **Fig. 3.16**.

The majority of data time series were received directly from responsible water sector institutions in Jordan and their respective monitoring programs. A considerable amount of data quality control and pre-processing on the original time series was necessary in order to prepare a consistent data basis, appropriate for the model application.

The resultant system understanding was expressed in a holistic water balance scheme to allow a general review and easy discussion with the Jordanian water sector professionals.

For the model construction the WEAP21 Water Evaluation and Planning (WEAP) tool was used. The applied model construction approach in this study can be understood as conceptual modelling, as the structure was specified before the modelling runs were undertaken, and some of the model parameters had to be estimated through calibration against observed data (Wagener & Wheater, 2006).

After the model was satisfactorily conceptualized, calibrated and validated, it was employed within an exemplary scenario planning exercise. For the latter, national water strategy objectives and action plans were analysed and used as normative guideline to craft a set of planning alternatives with direct relation to the local challenges. A concluding assessment compares the performance of the local scenarios to the national strategic objectives on the basis of a practically tailored indicator selection. Eventually, the applied modelling framework presented here, has the objective to provide decision makers with interdisciplinary and integrated planning support that is equally based on national IWRM policies as well as on sound science and thus with an instrument to progress towards operational IWRM. Furthermore, model development and scenario planning provided continuous input to the parallel development of the knowledge management approach, both in form of design requirements, as well as in modelling information to document and share.

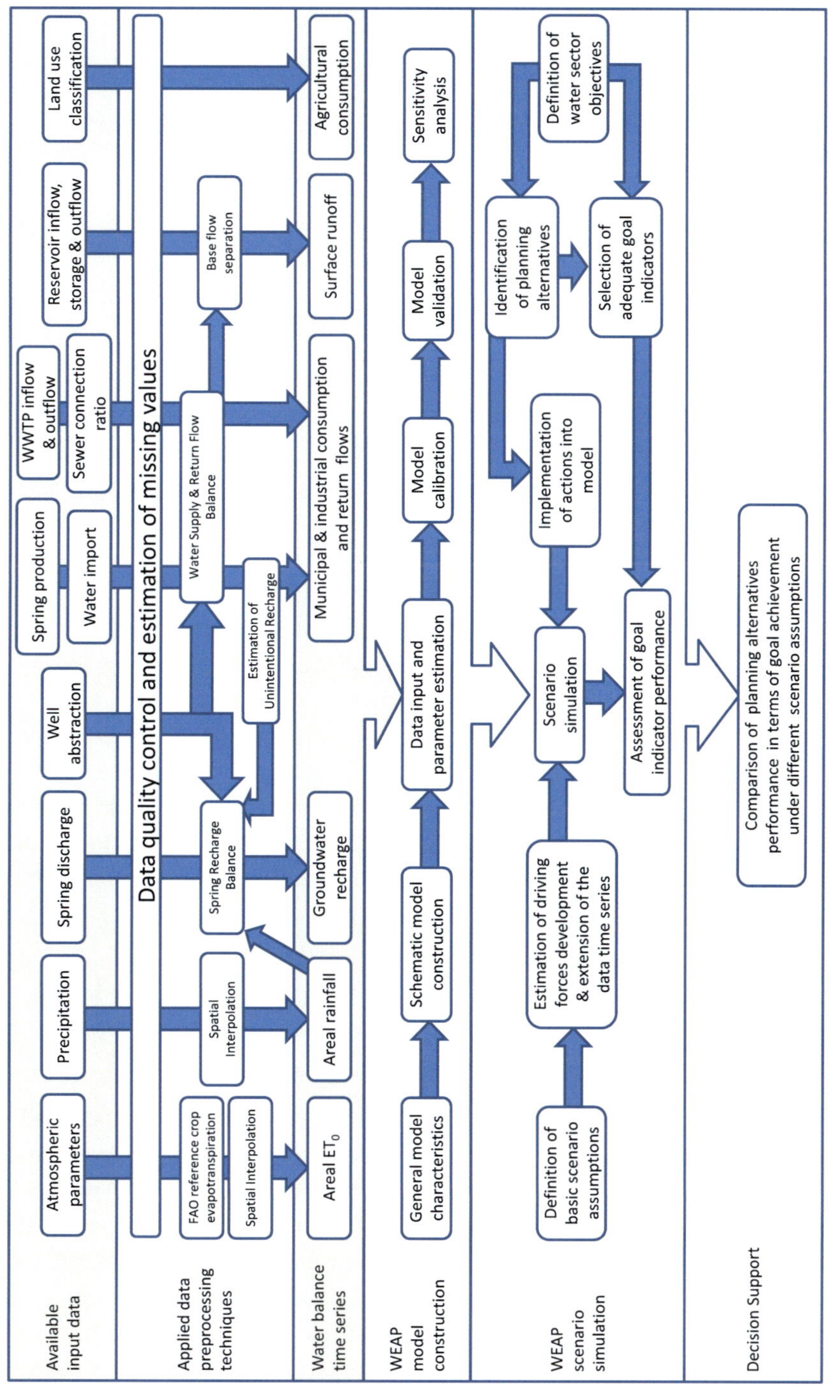

Fig. 3.16: Detailed workflow of the Wadi Shueib IWRM modelling and scenario planning study.

3.3.2. The WEAP21 Water Evaluation and Planning Software

The **W**ater **E**valuation **and P**lanning (WEAP) software tool was initially developed in 1988 in cooperation with the United States branch of the Stockholm Environment Institute (SEI-US) where it is still developed and maintained today. The recent version is officially labelled WEAP21, thereby giving credit to the major advances in comparison to the original model tool in terms of user interface, solution algorithms and the inclusion of several sub-modules. Stated design objectives of WEAP21 are to provide a water resources planning tool with hydrologic modelling, as well as management elements and a scenario planning environment (Yates et al., 2005).

The workflow in a WEAP21 application clearly follows a conceptual modelling approach by defining the model structure as "model schematic" prior to any model simulations. The basic structure is hereby composed as a system of water resources, demand sites and operational network elements as model nodes that are connected by flow vectors of various types (e.g. streams, canals, transmission links, return flows). Water balances are computed for every time step as linear programmed mass balance within lumped catchment nodes and the resultant fluxes are passed towards connected groundwater or river elements The conceptual model used for the balancing algorithm employs empirical functions to calculate evapotranspiration, surface runoff, interflow and deep percolation (Yates et al., 2005). By defining sub-catchments with differential parameterisations WEAP21 can also be employed to build at least semi-distributed catchment models (Yates & Strzepek, 1998). **Fig. 3.17** shows the input and output terms of the conceptual catchment node balance in WEAP21.

Groundwater can be modelled as simple aquifer bucket element with optional parameterisation of hydraulic conductivity, specific yield and river-aquifer interconnectivity. Within a sub-module WEAP21 also provides the possibility to link with a MODFLOW model, as well as to a Qual2K (water quality), and MABIA (crop water requirements and irrigation planning) model.

Following the catchment balance, for each time step the subsequent WEAP21 water allocation routine then uses the mass balance constants as received from the catchment nodes for a linear programming optimization of water fluxes between resources and demand sites with the objective function to maximize satisfaction of user defined demand priorities and flow requirements, by user defined supply priorities and quality demands. Demand site consumption and return flows are computed in the same fashion, so that at the end of each time step all received inflow to the system is either stored in the soil, an aquifer, a river, a tributary, a reservoir, or leaves the system by the end of that step (Yates et al., 2005).

WEAP21 models are built with a current accounts year, representing the actual water system situation and a reference time period that covers the planning horizon to be modelled. The reference period might receive an alternative and time-variant parameterisation in order to model future trends and developments in the system. Scenarios are inherited from the reference period (or from each other) and might again receive optional alternative parameterisations or additional system elements.

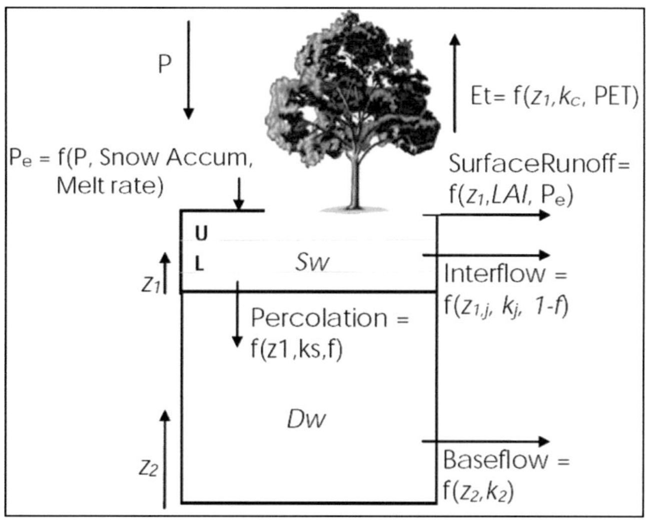

Fig. 3.17: WEAP21 conceptual water balance elements. The two bucket model tracks the storages Z_i by dividing forced rainfall into evapotranspiration (Et), runoff, interflow, percolation, and base flow for each defined fractional land use area with optional parameters for crop coefficients (k_c), leaf area index (LAI), water holding capacities (Sw&Dw), hydraulic conductivities ($k_{s\&j}$) and infiltration/runoff partitioning fraction (f) (Yates et al., 2005).

The WEAP21 model has been frequently applied in studies at various sites and with various spatial and temporal scales. The majority of studies used the software as instrument to assess the effects of changes in water infrastructure and water management under variable scenario assumptions, for example climate change or development scenarios (e.g. (Groves et al., 2008; Levite et al., 2003; Yates et al., 2009), or specifically for the Lower Jordan Valley region (Al-Omari et al., 2009; Alfarra et al., 2011; Hoff et al., 2011))[4].

---

[4] A comprehensive and regular updated list of publications in relation to WEAP applications can be found on the WEAP21 homepage: [http://www.weap21.org/] under "Publications" (*last accessed: 23/02/2012*)

3.3.3. Available Data

The recent history and the current state of the water system in the Wadi Shueib area, its interrelation with other sectors and the present challenges and deficits were assessed in a comprehensive review of the available data and information.

The focus of this modeling exercise was on the attempt of a holistic water sector model with scenario planning capability on the basis of the available data, there was no attempt to improve the modeling possibilities by conducting new measurement campaigns. Thus, the data collecting process that preceded the modeling effort, was built up of a comprehensive literature review on the study area and of the acquisition of a broad set of available data from the responsible water sector institutions (Ministry of Water and Irrigation, Water Authority of Jordan, Jordan Valley Authority) as well as from various related sector actors (Department of Statistics, Ministry of Agriculture, Ministry of Health) and related cooperation projects (GIZ, BGR, Dorsch Consult). This was accompanied by several consultations and interdisciplinary discussions with local sector experts from the governmental agencies, research facilities and development cooperation institutions as well as several field visits to the Wadi Shueib area.

According to Jordanian standards the Wadi Shueib catchment can be considered as rather well studied, and thus provided a useful basis for comprehensive water balance modelling. Nevertheless the available data and information shows essential gaps and discontinuities, especially when considering the temporal and spatial scale of the information (**Table 3.6**).

It has to be mentioned that missing data does of course not necessarily mean ultimate non-existence, however, the information was obviously not accessible for the decision maker and thus could not be regarded in any planning process.

Table 3.6: Overview of temporal and spatial scale of the available data for the water balance modelling in the Wadi Shueib catchment area (UFW stands for "Unaccounted for Water").

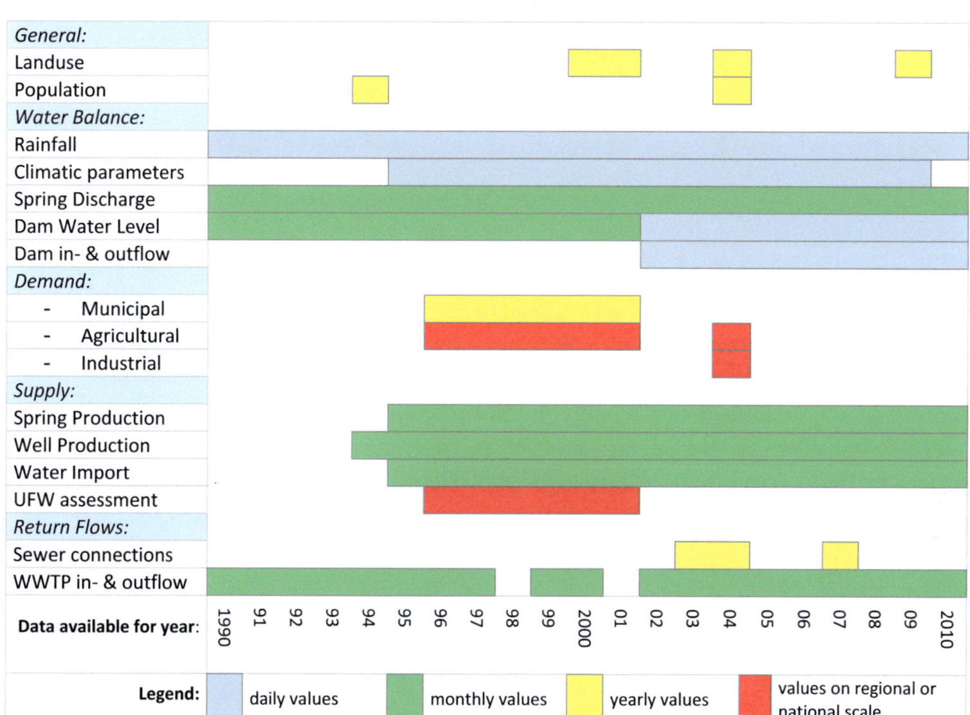

3.3.4. Data Quality Control

The majority of the available time series data for rainfall, atmospheric parameters, discharges, water production volumes and dam water balances were received from the Water Information System (WIS) database at the MWI or directly from employees at the MWI, WAJ or the Dams Department and originated from the institutions regular monitoring programs. Due to the lack of a working data quality control, much of the data had been in a rather bad qualitative state and required a fair amount of preliminary control, cleaning and pre-processing. Factual errors that could be assessed were:

- Obvious data input and computational errors
- Ambiguity and inconsistency of measurement techniques and routines
- Missing data points

Obvious data input and computational errors

Data input and computational errors can actually be very difficult to address as they generate random errors that only reveal themselves when they are of obvious size. Manual data inspection of the received data sets with the help of spread-sheet and graphing software revealed such obvious errors in abundance, mostly due to shifted decimal delimiters or added digits, which were manually corrected when possible. When the error was obvious but the true value not, the data point was completely deleted.

<u>Ambiguity and inconsistency of measurement techniques or routines</u>

Several of the received data sets were not self-explanatory in terms of applied measurement techniques, routines or even exact measurement locations. Furthermore, some data sets appeared to logically contradict observations at nearby locations. In such cases it was attempted to directly contact the responsible monitoring staff at the respective station or another knowledgeable person at the respective institution to clarify the exact data origins. This was, however, not always possible, due to the constant fluctuation of staff over the assessed monitoring period. In some cases the contacted staff at the institutions reported own uncertainty about data sets that have been received via external (often development cooperation) projects without adequate documentation. In extreme cases where no meta-information was obtainable at all, the data set had to be completely excluded from further analyses (e.g. the measured wadi discharge at station AM0008).

<u>Missing data points</u>

A high frequency of missing data points or even larger data gaps was found in some of the received time series. As the water balance modelling approach requires continuous time series at least for the calibration period, the gaps had to be filled. There are various statistical techniques for estimating missing data in time series promoted in the literature that range from simple interpolations using the sample means to very sophisticated and work-intensive ones such as multiple imputations. The suitability of a technique of course relates to the characteristics of the considered parameter as well as to the so called "missingness mechanisms" which accounts for the relation of the missing value(s) to the rest of the observed values (Scheffer, 2002).

The first step is always the identification of missing data. In the case of spatio-temporal continuous parameters (e.g. temperature), this can be achieved by simply checking the time series for gaps. In the case of intermittent parameters (e.g. precipitation or water flows), a data gap does not necessarily mean missing data but often that no measurable event occurred, thus the value should be set to zero.

The technique that was applied for the identification and estimation of missing data values in the Wadi Shueib case study was based on a single imputation process of missing values on the basis of observed values from the most similar neighbouring stations that were available from the MWI data. In order to define the most similar stations a similarity correlation between the stations was calculated as suggested by Lo Presti et al. (2010).

In order to preliminary reduce the number of possibly similar stations to the truly neighbouring ones, a search range threshold was chosen beforehand. In accordance to the findings of (Hubbard, 1994) and (Camargo & Hubbard, 1999) who examined spatial variability of daily weather variables in semi-arid USA, a 10 km search range was chosen for the similarity-search for rainfall gauges and of 50 km for evapotranspiration stations (**Fig. 3.18**).

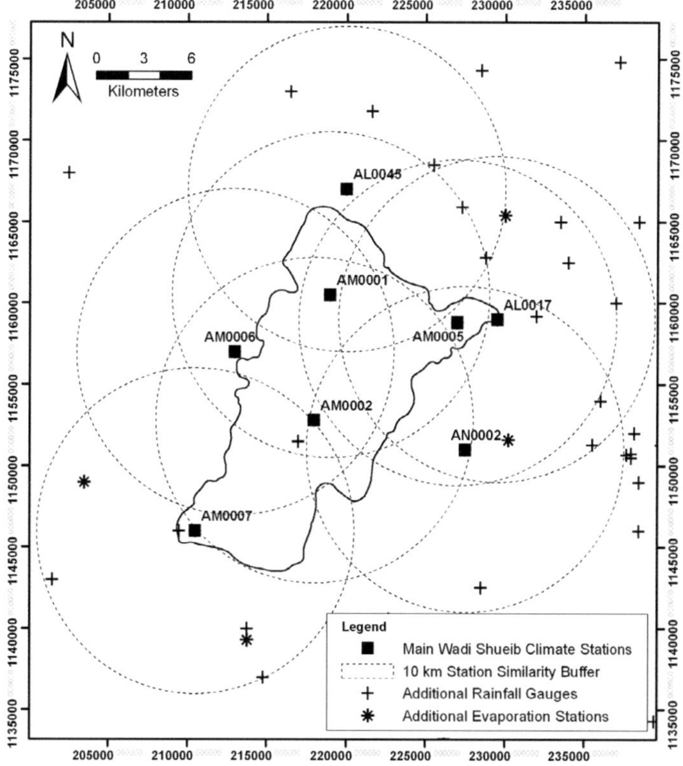

Fig. 3.18: In order to find neighbouring similar rainfall gauges for the assessment of missing data values a search range for of 10 km was used. For the evaporation stations the search range was extended to 50 km in accordance to the lesser variability of the temperature variable.

A total of 16 additional rainfall gauges and 4 additional evaporation stations fell into the search buffers and were subsequently examined on similarity.

As demonstrated by Lo Presti et al (2010) the similarity between stations can be assessed by means of the non-parametric Spearman correlation coefficient ρ. Giving respect to the abundance of tied values in the dataset the following formula for ρ was used:

$$\rho = \frac{\sum_{i=1}^{n} R_x(x_i)R_y(y_i) - \frac{n(n+1)^2}{4}}{\sqrt{\sum_{i=1}^{n} R_x(x_i)^2 - \frac{n(n+1)^2}{4}} \sqrt{\sum_{i=1}^{n} R_y(y_i)^2 - \frac{n(n+1)^2}{4}}}$$

where R_x is the rank of parameter x of one station, R_y the rank of the corresponding parameter y of another station and n is the total number of observations (Conover, 1971).

Calculated for every pair of stations, the Spearman coefficient was used to detect the most similar station, which was hence used to fill larger missing data gaps for the main Wadi Shueib climate stations.

3.3.5. Data Pre-processing

3.3.5.1. Areal Rainfall

The volume of received precipitation is the primary driving force of an area's water cycle. Thus, for water balance modelling it is necessary to find a satisfactory estimation of the total amount of precipitation at each time step.

The spatial interpolation (or regionalisation) of rainfall observations from a limited set of ground stations to estimate areal precipitation volumes and distributions is a fundamental hydrologic method and several techniques are currently in use. A good introductive overview to the history and current concepts on this matter can be found in Dingman (2008).

In order to select a suitable interpolation algorithm for the areal rainfall calculations in the Wadi Shueib catchment three typical techniques (Nearest Neighbour, Inverse Distance Weight, Ordinary Kriging) have been applied to the daily rainfall observations of the water year of 2004/05 for 10 rainfall gauges within the study area or its vicinity.

The Nearest Neighbour interpolation (also: *Thiessen polygon method*) is a conventional and often applied technique which simply assigns the record of the closest observation to the unsampled location, thus creating polygons of influence (Thiessen polygons; also: *Voronoi Tesselation*) around each gauge bordering at the halfway distance between all adjacent stations.

The Inverse Distance Weighted (IDW) interpolation estimates a parameter **u** at an unsampled location **x** as a linear combination of the surrounding observations weighted inversely proportional to the distance between observations (Shepard, 1968):

$$u(x) = \sum_{i=0}^{n} \frac{w_i(x)u_i}{\sum_{i=0}^{n} w_i(x)}$$

with

$$w_i = \frac{1}{d(x,x_i)^p}$$

where x is an unsampled location, x_i is an observation (sampled) location, n is the total number of observations, d is the metric distance between x and x_i and p is the power to which the inverse distance is weighted.

Ordinary Kriging (OK) interpolation is a geostatistical prediction technique that uses a semivariogram-function as a measure of dissimilarity to define the observation weights, instead of the Euclidian distance. It is largely used in the fields of hydrology and other earth sciences and often yields superior estimates when compared to the beforehand mentioned techniques (Tabios & Salas, 1985). Detailed discussions on various applied Kriging algorithms are found in Goovaerts (1997). As the significance of the Kriging method relies on a statistical evaluation of the sample data, it was questionable if the available sample size of 10 rainfall gauges would provide a sufficient basis for its application (see also Dirks et al. (1998) for a comparison of different areal rainfall interpolation techniques on small sample size basis).

The three interpolation sets were computed with the ArcGIS 9.3 *Spatial Analyst* module for the stations monthly and yearly rainfall heights in 2004/05. **Fig. 3.19** shows the maps of annual rainfall interpolated with the NN (left map), IDW (middle) and OK (right) interpolation.

In order to compare the estimations performance, each interpolation was run ten times, always excluding one station after another as reference observation point. The results were hence compared in respect to the calculated total amount of areal precipitation and the total mean square error (MSE) of prediction **z*** against the true observation **z** at each reference station with:

$$MSE = \frac{1}{n}\sum_{i=0}^{n}(z(x_i) - z^*(x_i))^2$$

where x_i is a sampled location (reference rain gauge), $z(x_i)$ is the observed rainfall at the sampled location, $z^*(x_i)$ is the predicted rainfall at the sampled location and n is the total number of reference sample locations.

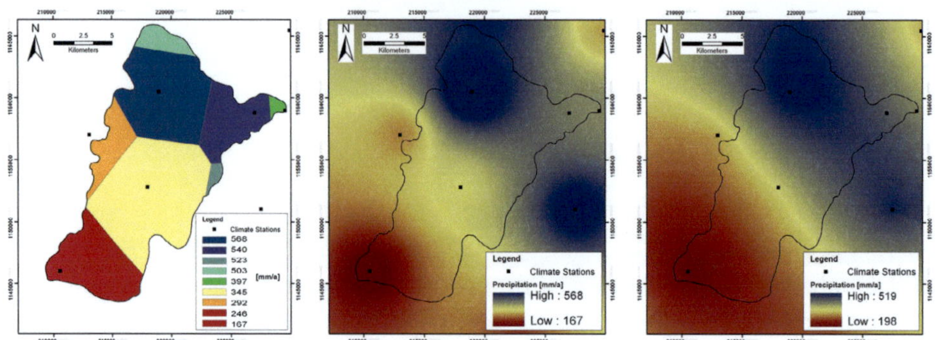

Fig. 3.19: Maps of annual areal precipitation in the Wadi Shueib catchment area for the water year 2004/05 (oct-sep), interpolated with the Nearest Neighbour (left), Inverse Distance (middle) and Ordinary Kriging (right) algorithms. The interpolation was based on ten rainfall stations in or near the study area (2 stations are not visible on the map display) (based on data records from the MWI).

Table 3.7 shows the results of the areal rainfall interpolation and the calculated total mean square error. The monthly variation of the areal rainfall estimations show that the IDW interpolation tends to predict the largest areal rainfall volumes whereas the OK tends to the lowest estimates. OK and NN show a closer correspondence with a total annual difference of 23 mm equivalent to 4.6 MCM, while IDW estimates 11.7 MCM more annual rainfall than NN or 16.2 MCM more than OK. The total MSE for each interpolation method shows that IDW produces larger prediction errors than NN and OK. It is interesting to note that NN and OK produce very similar prediction accuracy in the annual average but show greater variability in their performance for particular months. While NN for December 2004 showed the highest prediction error of all three methods, it outperformed the other algorithms for the majority of rainy months.

According to the results of the interpolation test, it appeared that the Nearest Neighbour-interpolation algorithm was the favourable approach for further application as it produced results and accuracies comparable to the Ordinary Kriging-algorithm but allowed a much easier and quicker application. To calculate the 20-year time series of monthly rainfall volumes for the Wadi Shueib catchment, the area of the Thiessen polygons intersecting the study area was taken as weighing factor for the stations influence on the area's rainfall. Thus, all calculations could be quickly done in spread-sheet software without the need for time-consuming GIS processing.

Table 3.7: Comparison of the monthly and annual areal rainfall interpolation results for the Nearest Neighbour (NN), Inverse Distance Weighted (IDW) and Ordinary Kriging (OK) algorithms and the total mean square error (MSE) of the prediction.

	Areal Rainfall [mm]			MSE		
Month	**NN**	**IDW**	**OK**	**NN**	**IDW**	**OK**
Oct 04	0.0	0.0	0.2	7	10	10
Nov 04	121.0	138.2	114.0	155	265	194
Dec 04	37.0	39.9	37.8	806	775	507
Jan 05	137.0	155.4	122.8	151	525	378
Feb 05	78.0	93.2	69.6	355	540	339
Mar 05	35.0	39.0	38.5	278	444	316
Apr 05	8.0	8.0	9.6	8	8	4
May 05	3.0	3.9	3.5	2	1	3
Jun 05	0.0	0.0	0.0	0	0	0
Jul 05	0.0	0.0	0.0	0	0	0
Aug 05	0.0	0.0	0.0	0	0	0
Sep 05	0.0	0.0	0.0	0	0	0
Total:	**419**	**478**	**396**	**Avg.: 147**	**214**	**146**

Table 3.8: Comparison of the results from different interpolation algorithm and given literature values for annual areal rainfall in the Wadi Shueib.

Interpolation	Average Area Precipitation	
	[mm]	**[MCM]**
Nearest Neighbour (with Thiessen polygons)	396.0	78.3
Inverse Distance Weighted (p= 2)	393.5	77.8
Ordinary Kriging	391.6	77.4
Ta'any (1992)	387.4	76.6
NWMP (2004)	395.0	78.1

3.3.5.2. *Areal Evapotranspiration*

Evapotranspiration (ET) as combination of evaporation and the biophysically controlled transpiration from vegetation is the major loss term in the water balance of semi-arid or arid catchment areas. Measuring ET, whether directly (e.g. with the eddy flux coefficient) or indirectly (e.g. with weighing lysimeters) is difficult and requires sophisticated and laborious technical equipment which is most often not available within an appropriate range of a study area. Pan evaporation and piche evaporation measures, regularly conducted at climate stations give a measure of daily evaporation potential from open water bodies, but need a location-specific adjustment coefficient to be transferred to ET potentials (Allen et al., 1998; Dingman, 2008).

For longer observation periods, the mean annual ET on a catchment scale can be reasonably estimated from the water balance term with known precipitation and runoff volumes. On shorter periods, e.g. months, the temporal water storage within the catchment cannot be neglected and water balance approaches have to incorporate these parameters. To quantify the ET in catchments without available direct measurements a multitude of approaches have been proposed that estimate potential and actual ET by employing empirical, e.g. (Blaney & Criddle, 1950; Hargreaves & Samani, 1985; Thornthwaite, 1948; Turc, 1961), or physically based models, e.g. (Monteith, 1973; Penman, 1948; Priestley & Taylor, 1972). A very valuable overview can be found in (Allen et al., 1998). These models were often motivated by agricultural research on irrigation requirements and usually use a reference surface. Evapotranspiration rates of different surfaces (typically crops) are then related to the reference surface by crop coefficients.

Besides the type of crop, the ET potential is affected by various factors including radiation, air temperature, air humidity and wind speed as atmospheric drivers; surface and soil characteristics, slope, aspect and vegetation cover as location parameters, as well as anthropogenic influence factors. Consequently, there is a large pool of investigations that compare different models under the conditions of specific study areas, climatic conditions and spatial and temporal scales (e.g. Amatya et al., 1995; Chiew et al., 1995; Itenfisu et al., 2003). These studies often conclude to the favour of one or another ET model for a specific study area and it seems difficult to beforehand predict the performance of an approach for a certain area.

Internationally accepted models were developed during the 1990s by the Food and Agriculture Organization (FAO) that follow Penman-Monteith and relate to a standardized grass (or alfalfa) surface area. The currently favoured model in the international literature is the FAO Penman-Monteith Reference Crop ET model (also: FAO-56) by Allen et al. (1998) which relies solely on meteorological parameters to estimate the potential evapotranspiration for a defined grass area with an assumed crop height of 0.12 m, a fixed surface resistance of 70 s/m, an albedo of 0.23 and unrestricted water availability by:

$$ET_o = \frac{0.408\Delta(R_n - G) + \gamma \frac{900}{T+273} u_2(e_s - e_a)}{\Delta + \gamma(1 + 0.34u_2)}$$

where ET_0 is the reference evapotranspiration [mm/day], R_0 is the net radiation at the crop surface [MJ/m²day], G is the soil heat flux density [MJ/m²day], T is the mean daily air temperature at 2m height [°C], u_2 is the wind speed at 2m height [m/s], e_s is the saturation vapour pressure [kPa], e_a is the actual vapour pressure [kPa], Δ is the slope of the vapour pressure curve [kPa/°C] and γ is the psychrometic constant [kPa/°C] which is given for average atmospheric conditions by:

$$\gamma = 0.665 \times 10^{-3} P$$

where P is the atmospheric pressure [kPa].

In order to transfer ET_0 from the reference grass area to other crops with different water requirements an empirically obtained crop coefficient is employed:

$$ET_c = K_c ET_0$$

where ET_c is the regarded crop evapotranspiration [mm/d], K_c is a crop specific coefficient, ET_0 reference crop evapotranspiration [mm/d].

The guideline for the computation of the FAO-56 PM ET_0 provides various suggestions on how to estimate missing parameters, as well as suggested crop coefficients for different crops, climates and growth stages (Allen et al., 1998).

For the Wadi Shueib Catchment, the potential evapotranspiration after FAO-56 was calculated for the records from two climate stations AL0035 to north and AM0007 at the Dam in South Shuna for the period from October 1995 to September 2009. The stations record daily measurements of minimum, maximum and averages of air temperature, pressure, humidity and radiation as well as average wind speed and Class-B-Pan evaporation measurements. **Fig. 3.20(a)** shows the 14-year averages of the calculated monthly ET0 and observed Pan evaporation records. The calculated ET_0 suggest a linear correlation between the two stations when observed at a monthly scale (**Fig. 3.20b**), whereas for the Pan evaporation records less linearity is observed especially in the summer months (**Fig. 3.20c**).

With regard to the limited set of available climate stations recording the necessary atmospheric parameters, a spatial interpolation approach as conducted for the areal rainfall interpolations (cf. Chap. 3.3.5.1) would result in two Thiessen polygons for the catchment area. But it would give superior influence to the South Shunah station (AM0007), resulting in a probable overestimation of evapotranspiration potential. Thus, to better represent the natural climatic separation of the study area into the Mediterranean upper catchment area and the semi-arid lower altitudes, the Wadi Shueib catchment was divided into two representative parts that would allow a differentiated assessment of the potential areal evapotranspiration losses in the water balances. The separation border was set in relation to the elevation range of the study area (~ –200 m bmsl to +1220 m amsl) at the 460m elevation isoline. **Fig. 3.21** shows the resultant upper (~121 km²) and lower (~77 km²) Wadi Shueib catchment area and the division of the Thiessen polygons for calculating the monthly areal rainfall in the upper and lower part. The two climate stations AL00035 to the north and AM0007 at the Dam in South Shuna were assumed as representative for the upper and lower part. **Fig. 3.22** shows the resultant time series from 1990-2009 for areal precipitation and areal reference evapotranspiration for the two catchment areas that were eventually exported as coma separated value (csv) files to be read by the WEAP software.

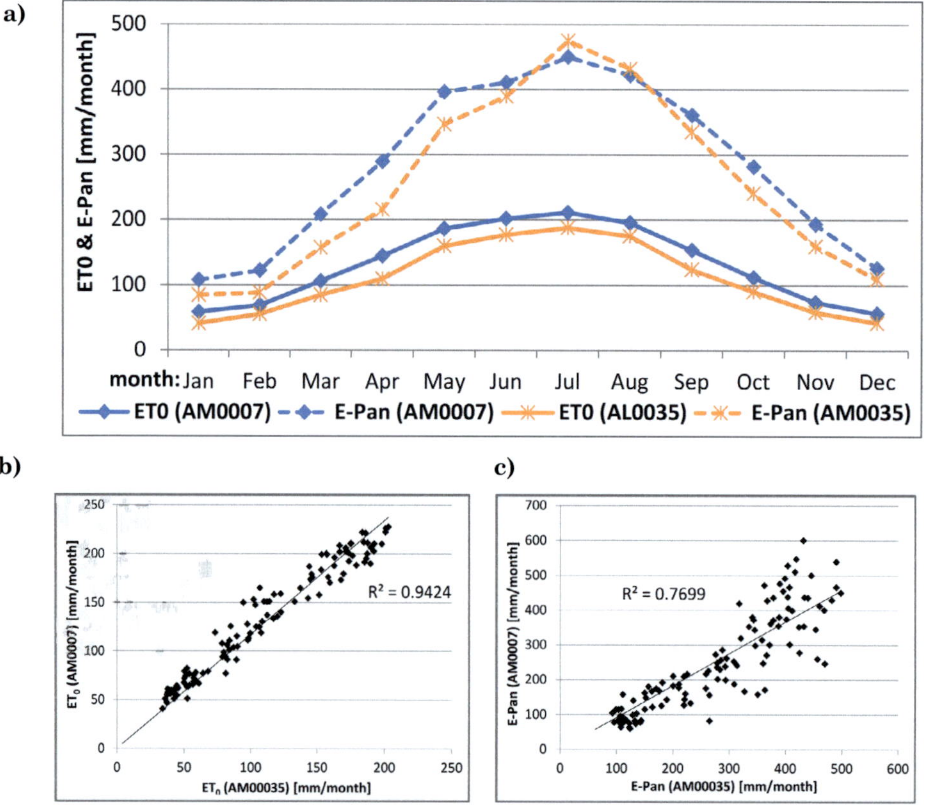

Fig. 3.20: (a) Average monthly ET0 and recorded Pan evaporation at the climate stations AL0035 and AM0007. (b)+(c) Scatterplots of the calculated monthly ET0 (b) and recorded Pan evaporation (c) at the two stations.

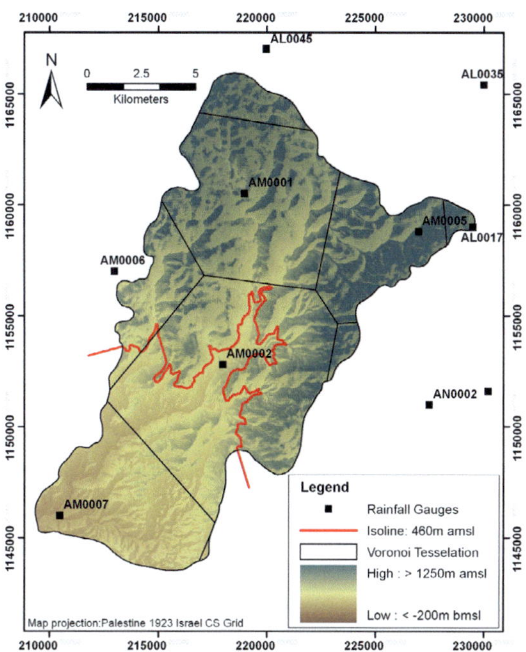

Fig. 3.21: Separation of the Wadi Shueib Catchment into an upper and lower part along the 460 m isoline. The Thiessen-Polygon areas around the rainfall stations used were accordingly divided for calculating the monthly areal rainfall in the upper and lower part of the Wadi Shueib Catchment.

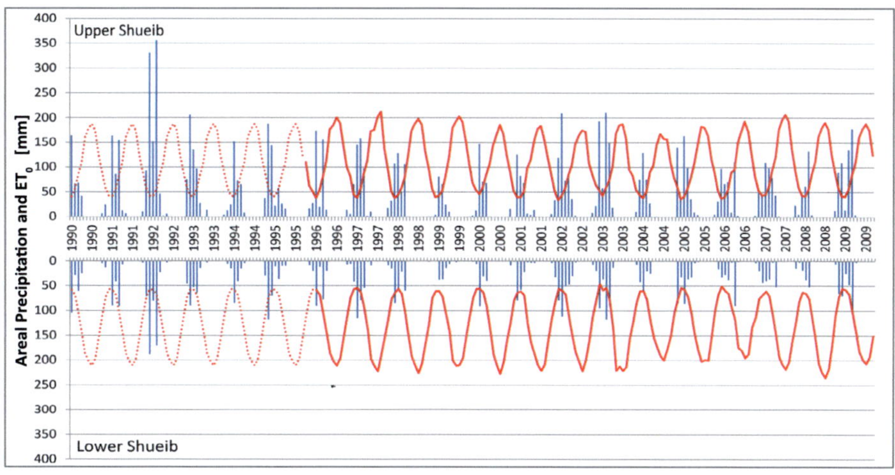

Fig. 3.22: Areal precipitation and reference evapotranspiration time series for the upper and lower part of the Wadi Shueib catchment.

3.3.5.3. Surface Runoff

Even though the Wadi Shueib WEAP model was not constructed to accurately simulate rainfall-runoff processes, the fraction of rainfall that generates direct surface runoff is nevertheless a basic model parameter as it is calculated as catchment-node outflow unavailable for the areal evapotranspiration loss. When the parameter is not set, the full amount of monthly rainfall is accessible to the potential reference evapotranspiration time series before infiltration or runoff volumes are subtracted, which predictably leads to an overestimation of evapotranspiration losses.

A simple approximation of the direct runoff can be deducted by subtracting the average monthly discharge records of the springs and waste water treatment plants from the recorded inflow volumes at the Shueib Dam. This approach, however, tends to underestimate the rising of the base flow after rainfall events. Ta'any (1992) conducted an analysis of the rainfall storms of the period from 1971 to 1991 by applying the Curve Number (CN) procedure of the United States Soil Conservation Service (US-SCS, 1972) to calculate the direct runoff- and infiltration ratios for the Wadi Shueib Catchment. According to his studies, the average volume of annual direct runoff amounts to 5.3 % of the annual rainfall. Due to the karst topography of the area the CN procedure might well underestimate the infiltration rate (Cazier & Hawkins, 1984).

Thus, to get a better estimation of the direct runoff for the actual model period a graphical hydrograph analysis was conducted on the basis of the daily inflow measurements at the Wadi Shueib Dam that were available for the period after March 2002.

The graphical base flow separation has already a long history in the field of hydrology (see Tallaksen (1995) for a good review on the development).

State-of-the-art techniques make use of today's processing power to apply multiple recession curve matching or use recursive digital filters to separate the high-frequency flood flow from the low-frequency base flow signal (Arnold & Allen, 1999).

Many of the proposed methods document high accuracy in the related literature but at the same time tend to require discharge records of high temporal resolution (hours or minutes), as well as an intensive work load for the modeller. The chosen method for this study on the other hand did not only need to promise robust results but also allow for the use of the available daily measurements. For this purpose the WHAT module (Web based Hydrograph Analysis Tool) by Lim et al. (2005) was found to provide an useful approach with satisfactory results.

The WHAT application offers three base flow separation modules: Local Minimum, BFLOW-filter and Eckhardt-Filter. The modules were tested against the available Wadi Shueib Reservoir hydrograph for the water year 2004/05 and the results were manually checked against the monthly discharge records of the springs and the waste water treatment plants. The Local Minimum module overestimated the base flow in large parts, thus the Eckhardt-Filter module was chosen for further application which uses a digital filter of the form (Eckhardt, 2005):

$$B_i = \frac{(1 - BFI_{max})\alpha + B_{i-1} + (1 - \alpha)BFI_{max}Q_i}{1 - \alpha BFI_{max}}$$

where B_i is the filtered base flow at the time step i, BFI_{max} is the maximum value of long term ratio of base flow to total streamflow, α is the filter parameter and Q_i is the total streamflow at the time step i.

General values for α (0.925) and BFI_{max} (0.8 for perennial streams with porous aquifers, 0.5 for ephemeral streams with porous aquifers, 0.25 for perennial streams with hard rock aquifers) are proposed by Eckhardt (2005). However, it was found that better results were obtained when estimating these parameters beforehand with the WHAT BFI_{max}-GA module (BFI_{max} Genetic Algorithm-Analyzer module) that applies an automated recession curve analysis to obtain a best fit to the given discharge data (Lim et al., 2010). **Table 3.9** shows the optimized values for BFI_{max} and α for the three water years of 2002/03, 2004/05 and 2005/06 accounting for wet, normal and dry water year conditions in the catchment. The model accuracy can be assessed by the Nash-Sutcliffe efficiency coefficient E (Nash & Sutcliffe, 1970), which is frequently used in hydrological modelling and basically sums the absolute squared differences between the modelled and the observed values normalized by the variance of the observed values for the observation period with:

$$E = 1 - \frac{\sum_{i=1}^{n}(Q_o^i - Q_m^i)^2}{\sum_{i=1}^{n}(Q_o^i - \bar{Q}_o^i)^2}$$

where Q_o^i is the observed and Q_m^i is the modelled discharge at the time step i. E can range from -1 to 1, the latter one being a perfect fit, whereas values below zero indicate that the mean value of the observed time series would have been a better predictor. With regard to the approach of finding best fit parameters for a complete water year and not for single runoff pulses, the predicted values appear to provide satisfactory results especially for the normal water and dry year conditions.

Table 3.9: Optimized Filter Parameter and BFI_{max} for the Eckhardt Digital Filter base flow separation module of WHAT for the Shueib Dam inflow hydrograph for three water years. The optimized values were obtained with the WHAT BFI_{max}-GA module.

Water year	Hydrologic conditions	Optimized Filter α	Optimized BFI_{max}	Nash-Sutcliffe coefficient
2002/03	Wet	0.9617	0.3835	0.6200
2004/05	Normal	0.9980	0.7139	0.8773
2005/06	Dry	0.9899	0.7944	0.7564

For the water year 2002/03 the model prediction tended to overestimate the base flow component during the intensive rain storms in March 2003. The optimized parameters were subsequently used to run a base flow separation with the Eckhardt-Filter module. The results of the base flow separation for the water year 2005/06 are shown in **Fig. 3.23**.

In order to deduct the runoff fraction parameter for the WEAP model, the modelled daily direct runoff values set in relation to the estimations of areal rainfall (cf. Chap. 3.3.5.1) to obtain a monthly rainfall-runoff ratio. The results are given in **Table 3.10**.

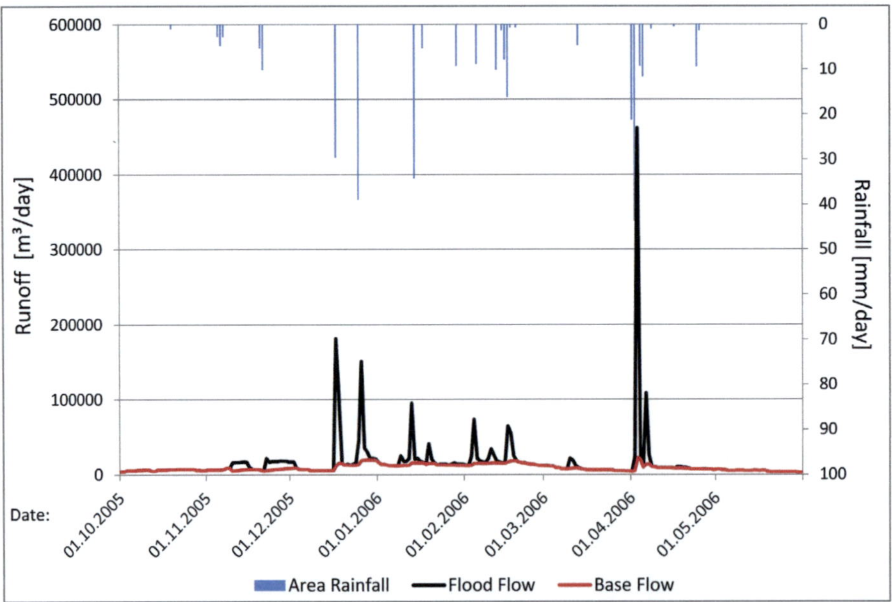

Fig. 3.23: Base flow separation with the WHAT Eckhardt Filter module for the water year 2005/06 and the calculated catchment areal precipitation.

Table 3.10: Estimated monthly rainfall-runoff for three water years.

Month	Areal Precipitation [m³]	Observed inflow at Reservoir [m³]	Estimated direct runoff [m³]	Direct runoff ratio [%]
Oct-02	311,850	179,453	0	0.0%
Nov-02	4,754,772	226,368	46,335	1.0%
Dec-02	25,966,710	1,928,448	1,464,176	5.6%
Jan-03	10,117,602	688,694	339,483	3.4%
Feb-03	22,434,390	1,424,477	965,931	4.3%
Mar-03	31,047,588	5,217,264	4,065,115	13.1%
Apr-03	2,617,956	3,632,688	1,853,709	70.8%
May-03	0	555,120	48,091	0.0%
Oct-04	0	172,195	0	0.0%
Nov-04	21,019,284	1,063,411	896,186	4.3%
Dec-04	5,734,872	476,410	208,809	3.6%
Jan-05	20,679,712	881,626	579,291	2.8%
Feb-05	14,065,920	902,016	568,355	4.0%
Mar-05	6,040,188	535,248	169,898	2.8%
Apr-05	1,532,916	280,800	24,811	1.6%
Oct-05	208,692	194,400	0	0.0%
Nov-05	5,118,300	370,224	165,543	3.2%
Dec-05	13,586,760	821,664	481,654	3.5%
Jan-06	7,847,136	588,816	173,194	2.2%
Feb-06	10,942,668	633,744	220,360	2.0%
Mar-06	926,640	256,608	29,817	3.2%
Apr-06	19,460,232	848,880	574,931	3.0%

The annual average of direct runoff calculated with the WHAT module are 2.8 %
for the dry year, 3.5 % for the normal year and 9.0 % for the wet water year
conditions. With exception of the very high runoff fraction after three months of
heavy rain in April 2003, the estimated ratios lie within the range that was also
found by Ta'any's (1992) application of the Curve Number analysis on the
rainstorm events between 1971 and 1991 who calculated annual runoff ratios in
the range of 1.2 % to 15.4 % with an average of 5.3 %. Furthermore, according to
the daily model results it appears that direct runoff is only observed at the Shueib
Reservoir after precipitation events of at least 1.35 MCM (or 7 mm) of areal
rainfall. The results for all years provided better estimations when compared to a
constant base flow approach and thus were adapted for the Wadi Shueib WEAP
model.

3.3.5.4. *Groundwater Recharge*

In union with the surface runoff fraction, the percentage of groundwater recharge from precipitation is an elementary allocation factor for the WEAP balancing module. Due to the complex geological setting and the lack of available observation wells an attempt for a groundwater flow model has never been undertaken in Wadi Shueib study area. Some authors have attempted generalized estimates of the groundwater recharge in the area but their approximations show a significant variance. Most effort has probably been conducted by Ta'any (1992) who calculated an average groundwater recharge of 12% from precipitation by means of a catchment wide water balance. However, he might have underestimated the recharge component by using the SCS curve number method (cf. Chap. 3.3.5.3). Zagana et al. (2007) applied a water-soil-balance model locally for the records of the As-Salt station and thereby estimated annual mean recharge rates of about 80 mm or 16 % of annual recharge for the area. In the continuous regional groundwater water model sustained by the Ministry of Water and Irrigation recharge rates of about 21 % of annual rainfall are assumed for the Jordan Valley Side Wadis (MWI, 2004).

So as to find a usable estimate of the monthly groundwater recharge rate for the model period a water budget balancing of the recorded monthly spring discharges and the areal rainfall in the recharge areas was undertaken. Due to the coarse temporal resolution of the spring discharge (monthly samples) records and the lack of soil moisture data, it was necessary to rely on annual balance calculations which needed subsequent disaggregation to approximate monthly recharge rates.

In order to define an area with best boundaries for in- and outflows, the groundwater contribution zone (GWCZ) of the four main springs in the Wadi Shueib area was selected, as delineated by the BGR&MWI (2010) in the framework of the spring protection zone project (**Fig. 3.24**).

According to Singhal & Gupta (2010) the balance for groundwater recharge for a given time period can be written as:

$$R_p = O_w + O_e + O_{gw} + ET - R_i - R_t - I_r - I_{gw} \pm \Delta S \pm n$$

where R_p is the recharge from precipitation, O_w is groundwater draft, O_e is effluent to rivers, O_{gw} is outflow to other groundwater units, ET is evapotranspiration, R_i is the recharge from field irrigation and canal seepage, R_t is the recharge from tanks, I_r is influent from rivers, I_{gw} is inflow from adjacent groundwater units, ΔS is the change in groundwater storage and n are undetermined elements of the balance and errors in estimations.

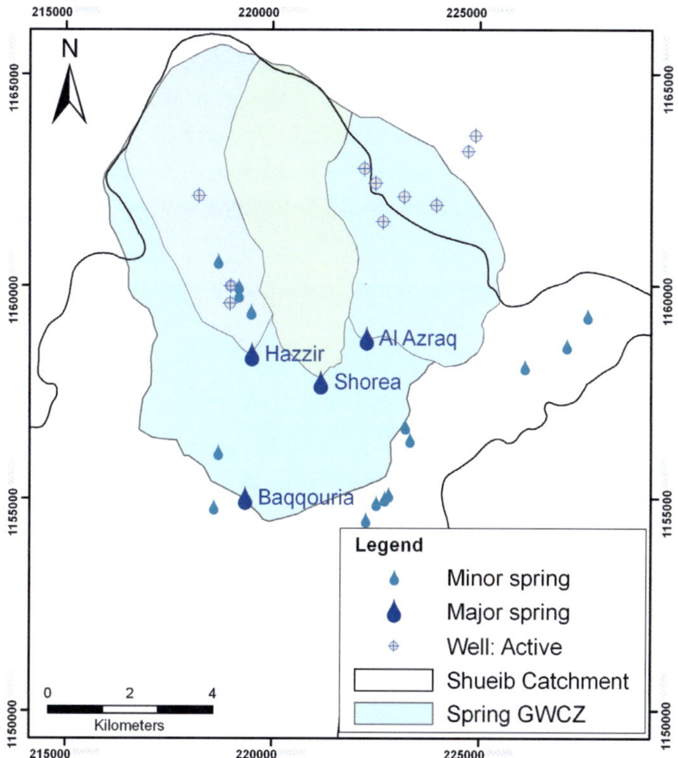

Fig. 3.24: Groundwater Contribution Zones (GWCZ) of the four major springs in the Wadi Shueib catchment and the locations of springs and wells in the area (modified after MWI&BGR, 2010).

With respect to the delineation of the Wadi Shueib springs GWCZ it was assumed that no groundwater inflow from adjacent units occurs and that all groundwater outflows are represented by the recorded spring discharges and well abstraction, whereas deep percolation towards the Kurnub (K1/K2) group was neglected.

According to Werz (2006) the average thickness of the unsaturated zone in the study area is assumed to range from 50 m (Hummar) to 60 m (Naur) and 75 m (Wadi As-Sir). Thus, direct evapotranspiration losses from the groundwater table were neglected.

Riverbed inflow was assessed together with the rainfall recharge volume as the wadi stream flow in the area is predominantly fed by precipitation runoff and spring discharge (although some feedback from spring discharge to the groundwater is of course possible, yet had to be neglected).

As agricultural activity in the area is reported as predominantly rainfed (Storz, 2004), additional infiltration from irrigation water is neglected. Seepage from the water supply and the sewer system and septic tanks was taken into account as

Unintentional Recharge (UR) volume and was estimated by a water supply and return flow balance (cf. Chap. 3.3.5.5).

Groundwater storage in the GWCZ was estimated based on Ta'any's (1992) conclusions from his spring hydrograph and flow-duration curve analyses. The summer storage volumes given therein were grouped according to wet, dry and normal water year conditions and applied for respective water years in this studies model period.

Based on the given assumptions a modified recharge balance for the Wadi Shueib springs GWCZ was applied with:

$$R_p = O_s + O_w - R_{UR} \pm \Delta S \pm n$$

where R_p is the recharge from precipitation, O_s is the spring discharge, O_w is the well abstraction, R_{UR} is the unintentional recharge from water supply pipelines, sewer canals and septic tanks, ΔS is the change in groundwater storage and n is the error term.

Table 3.11 shows the resultant groundwater recharge balance for the Wadi Shueib Springs groundwater contribution zones for the water years 1995/96 until 2008/2009.

According to the annual balance the mean groundwater recharge volume in the area is 9.9 MCM per year or 21 % of the areal precipitation over the delineated groundwater contribution zones. The estimated yearly recharge volumes range from 8.1 MCM in the very dry winter 1998/99 to 11.9 MCM in the wet winter of 2002/03. The fraction of precipitation contributing to groundwater recharge ranges from 16 % to 37 %, whereas it appears that higher relative recharge occurs during dry winters in comparison to normal and wet water years. This would be in agreement to the study of Jiries et al. (2010) who found a higher percentage of soil infiltration relative to total precipitation for drier water years in their irrigation efficiency studies in the Shueib area. Furthermore, the estimated annual recharge rates show a good agreement with the recharge figures used in the regional groundwater modelling at the Ministry of Water and Irrigation (MWI, 2004), but is considerably higher than the average estimates of Zagana et al. (2007) and Ta'any (1992).

Possible errors due to the boundary assumptions might be especially the assumed absence of further subsurface groundwater outflows from the GWCZ. When taken into account, average recharge fractions could well reach up to 30 % of annual recharge as found by Alkhoury (2011) for the adjacent Wadi Kafrein area.

Table 3.11: Annual recharge balance for the Wadi Shueib springs groundwater contribution zones for the period from 1995/96 to 2008/09.

Water year	Areal rainfall	Spring discharge O_s	Well abstraction O_w	Unintended recharge R_{UR}	Change in storage ΔS	Recharge from precipitation R_p	
				[MCM]			[%]
1995/96	46.5	10.2	2	1.4	-0.7	10.8	23%
1996/97	53.4	10.3	2	1.4	0.2	10.9	20%
1997/98	50.4	10.4	1.8	1.4	0	10.8	21%
1998/99	22.1	7.1	1.9	0.9	-3.6	8.1	37%
1999/00	33.6	8	2.2	1	-1.4	9.2	27%
2000/01	37.8	8.1	2.1	0.9	-0.7	9.3	25%
2001/02	63.1	11.2	1.3	1.2	1.7	11.3	18%
2002/03	74.8	11.9	1.4	1.4	3.6	11.9	16%
2003/04	38.7	9.8	1.5	1.4	-2.9	9.9	26%
2004/05	56.3	9.7	1.7	1.4	2.1	10	18%
2005/06	46.3	9.1	1.6	1.4	0.5	9.3	20%
2006/07	53.9	9	1.4	1.4	0.5	9	17%
2007/08	34.5	8.1	1.4	1.3	-1.3	8.2	24%
2008/09	62.2	9.6	1	1.3	2.4	9.3	15%
Avg.	48.1	9.5	1.7	1.3	0	9.9	21%

3.3.5.5. *Water Consumption and Return Flows*

The natural water balance of the Wadi Shueib catchment is already highly influenced by human activities which were assessed in a preliminary study of the municipal, industrial and agricultural water supply, consumption and return flow processes in order to define the allocation principles for the water balance model. The water sector challenges in the study area are resulting in large parts from water losses in the water supply as well as the waste water system (cf. Chap. 3.2.8 and 3.2.9).

In order to account for the monthly water supply and return flow losses in the study area, the available records of springs and well production and water imports were balanced against the demographic figures of the supplied municipalities and the recorded inflows at the two waste water treatment plants for As-Salt and Mahis and Fuheis. With the losses from the water supply being:

$$UR_{FW} = (P_S + P_W + P_I - P_E) \times UFW_{ph}$$

where UR_{FW} is the volume of unintended recharge of freshwater from supply network losses at a demand site [m³/month], $P_S/P_W/P_I/P_E$ are the spring/well/import/export freshwater productions allocated to the demand site [m³/month] and UFW$_{ph}$ is the ratio of physical network losses from the unaccounted for water [dimensionless] which is estimated at 0.286 for the study area (MWI, 2004).

The volume of generated waste water was estimated from the recorded treatment plant inflows and BOD values, assuming an average BOD of 60 g per capita per day. The losses from the waste water return flows were assumed to occur primarily from cesspits and septic tanks and to correspond to the percentage of cesspits reported as being never emptied (cf. Chap. 3.2.8) thus giving:

$$UR_{WW} = \left[\frac{WWTP_{in}}{Pop_s}\right] \times Pop_{ns} \times CP_{leaking}$$

where UR$_{WW}$ is the volume of unintended recharge of waste water from return flow losses at a demand site [m³/capita/month], WWTP$_{in}$ is the recorded monthly waste water treatment plant inflow [m³/month], Pop$_s$/Popn$_s$ is the number of people served/not served by centralized waste water service [capita] and CP$_{leaking}$ is the reported percentage of cesspits never emptied [58 %].

Table 3.12 shows the results of the return flow balance for the two districts of As-Salt and Fuheis/Mahis. The municipal supply represents the total water supplied from the public network for domestic and non-domestic uses after the subtraction of the physical network losses (UR$_{FW}$). It includes water imports to the network via the Zai- and Dabouq-pipelines, but it does not include water bought from private vendors and thus could underestimate the actual water consumption. The annual per capita supply ranged from 33.9 m³ to 50.0 m³ (~93-137 l/c/d) in As-Salt and from 36.5 m³ to 52.8 m³ (~100-145 l/c/d) in the Fuheis/Mahis sub-district with the lowest supply during the little water crisis in 2001/02 after three consecutive drought years. The proportion of imported water amounts to 41 % of the supply for the As-Salt district and to 36 % for Fuheis and Mahis. The consumption loss ratio expresses the amount of supplied municipal water that is not recorded as waste water, e.g. used in private and commercial gardening or public areas, or is otherwise lost to evapotranspiration. The consumption losses amount to an average of 25-30 % but display a considerable variance between years, which is not unusual according to the literature (Tchobanoglous et al., 2003). The average amount of generated waste water per capita accounts to 80 l/capita/day in the As-Salt district, and 84 l/capita/day in Fuheis and Mahis, which amounts to about 93-94 % of the total domestic consumption. Within a household survey study on greywater reuse potential Jamrah et al. (2008) come to comparable values for the nearby Zarqa area where they found 93% of total domestic consumption was generated waste water. It is estimated that approximately 4-12 % of the generated waste water in the Wadi Shueib area is leaking into the underground

(UR$_{WW}$). Due to the lesser sewer connectivity the ratio is higher in the As-Salt sub-district compared to Fuheis and Mahis. As mentioned above, the UR$_{WW}$ in this model takes a conservative estimate on losses solely from cesspits and tanks. Possible leakages from sewer canals are hereby ascribed to the consumption loss term. Thus, the UR$_{WW}$ may also be underestimated and understood as lower bound on the loss parameter.

Table 3.12: Return flow balance for the two districts of As-Salt and Fuheis/Mahis for the period from 1998 to 2009.

	Sub-district As-Salt				Sub-district Fuheis/Mahis			
Year	Municipal supply [m³/cap]	Consump. loss [%]	WW$_G$ [m³/cap]	UR$_{WW}$ [%]	Municipal supply [m³/cap]	Consump. loss [%]	WW$_G$ [m³/cap]	UR$_{WW}$ [%]
1998	50.0	39%	30.0	8%	50.6	41%	20.2	8%
1999	42.6	45%	23.3	10%	40.1	37%	23.8	6%
2000	39.1	35%	24.7	9%	40.0	28%	27.9	6%
2001	33.9	23%	25.5	10%	36.5	35%	23.4	7%
2002	37.4	23%	27.7	9%	52.8	30%	33.4	5%
2003	39.3	22%	29.8	9%	46.7	25%	34.2	5%
2004	43.9	28%	31.4	9%	43.7	20%	33.2	5%
2005	46.4	31%	31.5	10%	41.1	14%	32.8	5%
2006	44.0	33%	29.5	11%	45.0	21%	33.5	5%
2007	43.2	30%	30.3	11%	46.7	22%	34.6	4%
2008	36.3	17%	30.1	12%	45.4	21%	35.0	4%
2009	38.6	16%	32.4	11%	47.2	12%	37.6	4%

3.3.6. Wadi Shueib Water Balance

In order to combine and review the results of the systematic pre-processing steps, the resultant system understanding was expressed in holistic water balances for different water year conditions. The conceptual balancing approach was straightforward in defining different compartments and their inflows and outflows. It is inevitable that certain flow paths or storage components stay highly uncertain. Elements without the availability of measurement data or directly related studies, and which also cannot be properly balanced from other components, have to be estimated from the experience of experts or the international literature. In the case of the Wadi Shueib water balance this primarily concerns several components of the groundwater flow, and here especially the sewer leakages and the resultant unintentional recharge, as well as potential lateral groundwater flows. Furthermore, the irrigation related evapotranspiration and return flows, as well as the amount of private water imports. Nonetheless, the international literature provides a multitude of examples where even highly uncertain water balancing approaches have been usefully applied in the water resources studies at catchment scales (e.g. Fleischbein et al., 2006; McCartney, 2000).

In order to allow general review and efficient discussion with project partners and Jordanian water sector professionals, the prepared balances where compiled in balance schemes that employ coloured flow arrows of variable size to express significance and uncertainty of the information. This visualization proved very useful in the communication during discussions on the water system understanding and helped significantly in acquiring expert opinions and estimates, thus reducing some prevailing uncertainties. A final result of these balance schemes is illustrated **Fig. 3.25**, which exemplarily summarizes the acquired knowledge of the Wadi Shueib water budget for the water year 2008/09. Within the SMART project, the structured visualization of the current state of knowledge could also be used to delineate areas with critical knowledge gaps and to help streamline further research activities.

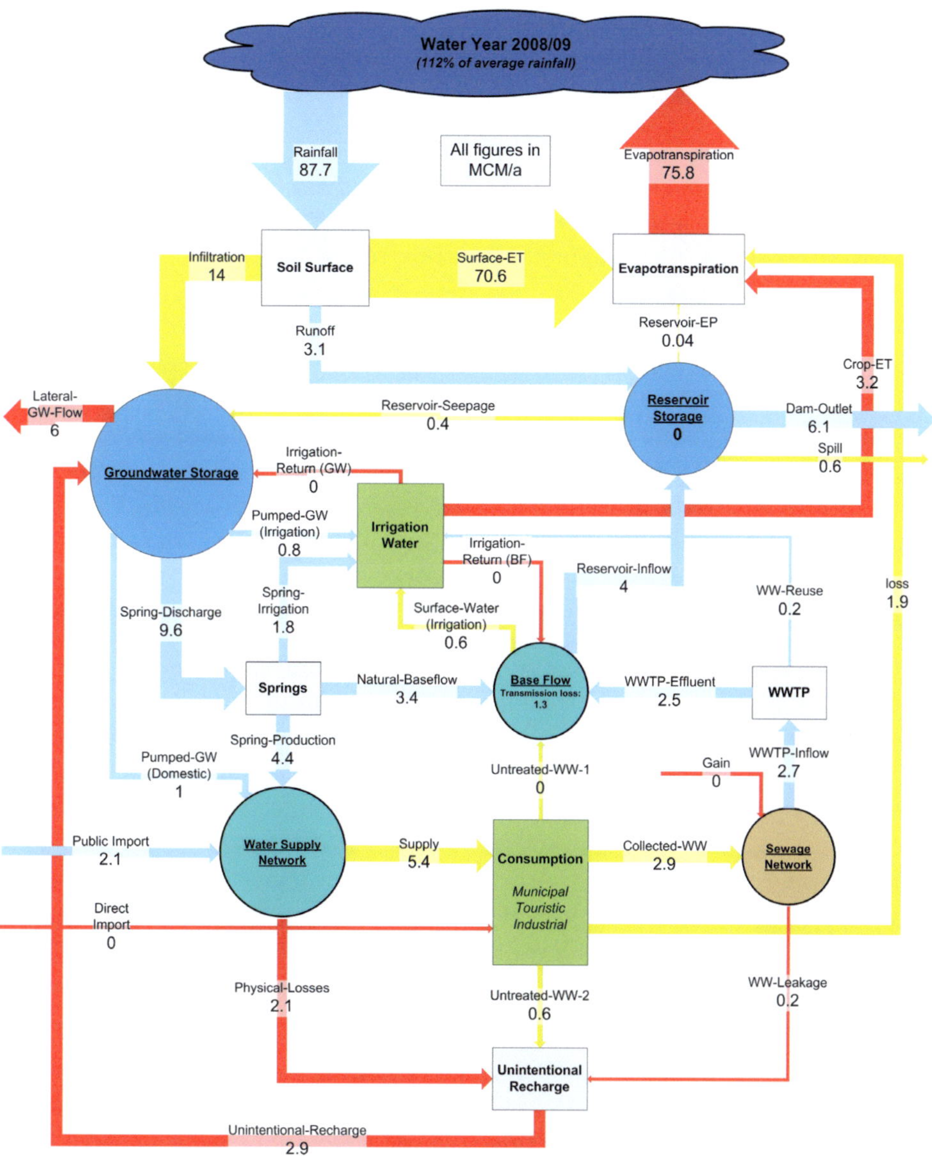

Fig. 3.25: Water Balance for Wadi Shueib compiled from the pre-processed data. All values are given in MCM for the water year 2008/09. The colour of the flow arrows indicates the confidence in the data quality: blue: good (measured); yellow: medium (balance derivate); red: low (expert or literature estimates or no data). Transmission losses (baseflow) are an outflow component.

3.3.7. Wadi Shueib WEAP-Model

3.3.7.1. General Model Characteristics

Spatial Scale

The spatial boundaries chosen for the IWRM-model are oriented after the Wadi Shueib Dam surface water catchment (cf. Chap. 3.2.2 - Fig. 3.2), which was obtained by applying the ArcMap©-*Watershed*-algorithm on a 1:20:000 digital topographic map of the area. For the groundwater recharge estimations, the subsurface catchment was assumed to be slightly extended to the north in comparison to the surface catchment, as it was defined in the Spring Protection Zone Study by the BGR and MWI (BGR & MWI, 2010). Furthermore, to integrate water imports and exports three additional external demand sites were added to the model, representing the communities that receive water supply from the Wadi Shueib springs and wells, as wells as the irrigated areas in the Jordan Valley that receive the Shueib Dam Water.

Temporal Scale

The decision on the modelling time steps has to depend primarily on the intended application. Daily and monthly time steps are common in water balance models on catchment scale. Shorter time steps become necessary when the model objectives focus on the physical representation of runoff or solute transport processes. It is important to note is that smaller time steps usually involve higher model complexity and data requirements, whereas the accuracy of the predictions is usually experienced higher in monthly models in comparison to daily models (Ye et al., 1997). For the purpose of yield-based water resources assessment, models with monthly time steps have shown to provide pragmatic and applicable approaches (Vandewiele & Ni-Lar-Win, 1998) and have also been successfully applied in arid to semi-arid regions (Moreda, 1999).

With regard to the available data and the objectives a monthly time step was chosen for the Wadi Shueib WEAP model with a calibration and validation period from 2001 until 2009. For the planning scenario simulations the time frame was expanded until 2025, regarding the actual MWI water planning horizon.

3.3.7.2. Schematic WEAP Model

The conceptual model of the Wadi Shueib water resources and supply structure is represented in the WEAP model schematic (**Fig. 3.26**). The modelled components are hereby represented as set of interconnected nodes of catchment areas, aquifers, supply and demand sites, waste water treatment and water storage facilities. During model simulation the nodes act as one-dimensional lumped parameter elements wherein the water budget calculations for each time step are performed. The linking vectors between the nodes represent runoff or infiltration allocations, natural stream flows and intentional water transmissions and return flows. Stream flows and diversions can be modelled as two-dimensional features for some calculations, given the reach lengths and related parameter-variance.

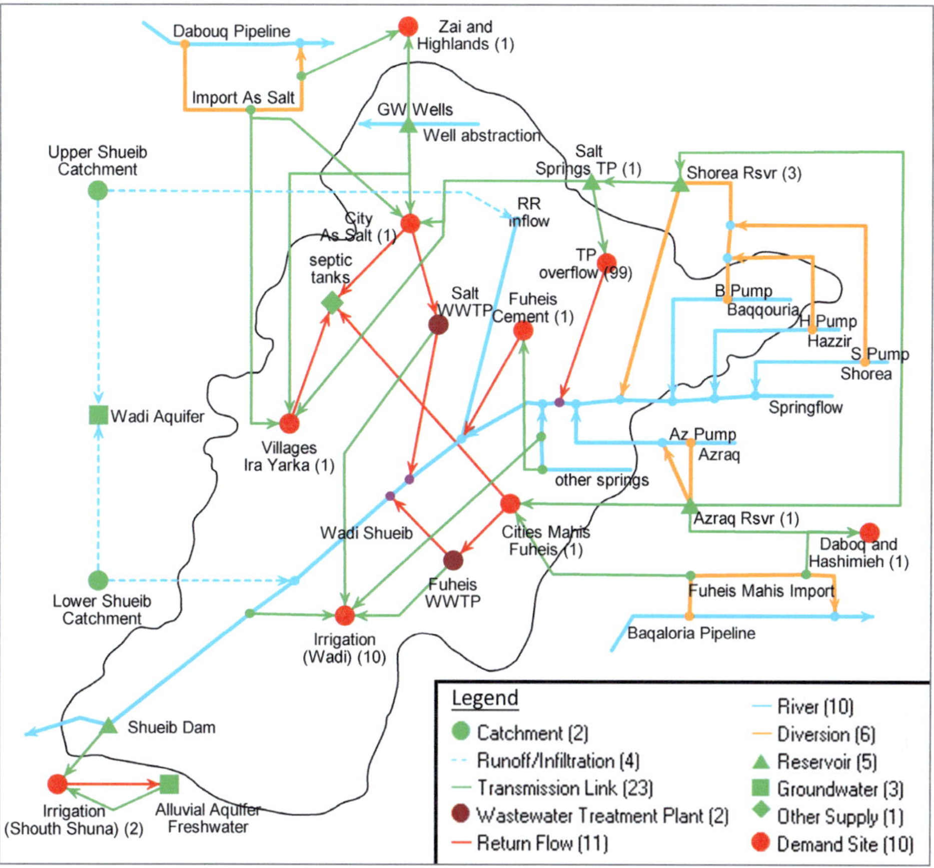

Fig. 3.26: Schematic representation of the Wadi Shueib WEAP-model. In order to provide a better overview some nodes are shifted from their true geographical location.

The Wadi Shueib catchment is represented as two separated sub-catchments (Upper: 121 km² and Lower: 77 km²) as discussed earlier (cf. Chap. 3.3.5.2) that drain received precipitation towards the main Wadi Shueib stream and the aquifer (cf. Chap 3.3.5.3 and Chap. 3.3.5.4). The hydrogeology in the catchment is modelled as single cretaceous aquifer complex (A1-A7) that receives recharge from rainfall as well as unintentional recharge from the water supply transmissions and waste water return flows (cf. Chap. 3.3.5.5) and feeds the various springs and wells, which are all represented as receiving streams. Springs and wells are pumped and allocated towards small reservoirs which either directly supply municipal and industrial demand sites or receive pre-treating in the Salt Springs Treatment Plant (cf. Chap. 3.2.8). The municipal demand sites also receive further supply by diversion from two pipelines running for Amman city (cf. Chap. 3.2.8). Demand site waste water is allocated according to the sites respective sewer connection ratio, either towards the waste water treatment plants or towards the septic tank node (cf. Chap. 3.2.8). Septic tanks and cesspits were modelled as a single receiving node without return flow allocation, their significant waste water leakage (58 %) occurring in the inflow flow links and the residue assumed to leave the system by pumping trucks to be treated elsewhere. Treated waste water from the plants is either discharged to the wadi channel or directly supplied to some agricultural plots in the vicinity of the facilities (~0.2 MCM/a). Further agricultural demand is supplied by the smaller springs as well as from the wadi runoff (~1.8 MCM/a) (cf. Chap.3.2.7). The wadi channel eventually drains into the Wadi Shueib Reservoir to be stored for downstream agricultural use or to be directly spilled towards the valley when the reservoir capacity (1.43 MCM) is reached (cf. Chap. 3.2.6.2). A second aquifer downstream of the catchment represents the alluvial deposits that are recharged through the Wadi Shueib dam seepage and abstracted by the local farmers in the Jordan Valley.

In order to define the allocation rules for the model the elements are given supply priorities, according to which sequence their demands are to be fulfilled. In the Wadi Shueib model, all municipal and industrial demand sites received the highest priority (1), whereas agricultural areas and reservoir volumes received lower values (2-99) in order to represent the water system arrangement.

The prepared time series of monthly data were linked to the respective schematic elements to define the discrete model input for each time step. Recorded flow volumes where hereby implemented within the respective transmission links as monthly variable maximum flow volume. Fixed ratios of water supply and return flow leakages were accordingly attributed to the respective transmission or return flow links. The demand sites within the catchment were given a fixed ideal supply as demand per unit in order to account for occurring supply shortages (100 l/cap/day for municipal and 0.5 MCM/km^2/a for agricultural demands).

After initial parameterization the WEAP model was calibrated to match simulated and observed storage volumes in the Wadi Shueib Reservoir on a monthly time step.

3.3.7.3. Model Calibration and Validation

Model calibration covers the process of determining best estimates for unknown or uncertain model parameters by comparing model simulations to observed records for a specific output or set of outputs and for a given time period and fixed conditions. Subsequent model validation uses the calibrated parameters to compare model outputs to observations for further time periods and conditions in order to demonstrate the models capability of making sufficiently accurate predictions in regard to the model objectives (Refsgaard, 1997). Evaluation of the model validity is usually done by statistical and graphical performance criteria for the simulation-observation comparison.

Calibration and validation procedures and respective evaluation criteria for watershed models have been proposed in abundance in the reviewed literature, whereof Xu & Singh (1998) as well as Moriasi et al. (2007) give very helpful comparative reviews. Typical steps of the model calibration involve the choice of a representative calibration period and calibration parameters, the choice of adequate evaluation technique(s), and the actual iterative calibration procedure.

Calibration might comprise one or more calibration parameters and can be conducted manually or (semi-) automatically by applying objective functions and iterative search algorithms to the parameters. Manual calibration still is the conventional approach especially when the parameter set is limited (Moriasi et al., 2007). Studies have also shown that a smaller number of calibration parameters usually provides more accurate model performance and at the same time increases the information content of each single parameter (Servat & Dezetter, 1993).

Moriasi et al. (2007) classified the quantitative evaluation criteria commonly employed in hydrologic models into three groups as:

(1) Standard Regression (e.g. Pearson correlation coefficient, coefficient of determination R^2),

(2) Error Index (e.g. (Root) Mean Square Error, Percent Bias) and

(3) Dimensionless measures (Nash-Sutcliffe Efficiency),

and recommended the use of a combination of dimensionless and error index statistics, as well as graphical evaluation criteria. The latter comprise the visual comparison of simulation and observation in order to evaluate the model performance (e.g. by hydrograph analysis or Percent Exceedance Probability curves).

The Wadi Shueib WEAP model was calibrated to match simulated and observed storage volumes in the Wadi Shueib Reservoir on a monthly time step. As characteristic calibration period a sequence of three consecutive water years with wet, dry and average conditions was chosen: 2002/03, 2003/04 and 2004/05 with

145 %, 75 % and 108 % of long term average areal precipitation. Once calibration was achieved for the 2002-2005 period, the model was put to a validity test by comparing observed and simulated reservoir storage for the years 2001/02 and 2005-2009 (*Split sample test*).

With the considerable amount of effort invest into pre-processing the time series for the model input, a preliminary calibration of various balance parameters against records of spring discharge and treatment plant outflows had already been conducted (cf. Chap. 3.3.5). The obtained a priori estimates of rainfall, evapotranspiration, infiltration and return flow volumes were thus assumed reasonable and treated as fixed time series input data. With the given records of reservoir outflow and pan evaporation measurements, the remaining calibration parameters governing the reservoir storage could be thus reduced to two parameters: the *transmission losses of surface runoff* in relation to flow volume and the *reservoir seepage* in relation to reservoir storage.

The calibration was conducted manually by iterative adjustment of the monthly parameter values after model runs. In order to stay within reasonable bounds and in relation to the input time series a set of threshold values were forced. Transmission losses were assumed to lie within the range of 0-35 % of storm runoff with regard to the estimations by Alkhoury (2011) for the adjacent Wadi Kafrein area. Furthermore they were not allowed to alter flows beyond the range of the estimated runoff fractions for dry, wet or average years (cf. Chap. 3.3.5.3). The reservoir seepage was assumed between 0 % - 27 % of stored volume, according to the highest estimation of reservoir seepage from the comparable Wadi Kafrein Dam (Nusier et al., 2002). The calibration procedure started with a minimum assumption of 0 % transmission loss which was iteratively raised by 1 % steps, while for every step the best estimate for seepage ratio was iterated. The initial steps were evaluated visually until a satisfactory preliminary fit was found, then the statistical evaluation parameters were used for further fine-tuning.

The model validity for the calibration and validation period was quantitatively evaluated by Nash-Sutcliffe Efficiency and Percent BIAS.

The Nash-Sutcliffe efficiency coefficient NSE (Nash & Sutcliffe, 1970) is an indicator for how well the observed versus simulated data fits the perfect match by summing the absolute squared differences between the modelled and the observed values normalized by the variance of the observed values for the observation period with:

$$NSE = 1 - \frac{\sum_{i=1}^{n}(P_o^i - P_m^i)^2}{\sum_{i=1}^{n}(P_o^i - \bar{P}_o^i)^2}$$

where P^i_o is the observed and P^i_m is the modelled parameter at the time step i. NSE can range from -1 to 1, the latter one being a perfect fit, whereas values below zero indicate that the mean value of the observed time series would have been a better predictor.

The Percent Bias PBIAS reveals the average tendency of over- or underestimation of the simulation. It has an optimal of 0.0% while positive values indicate model overestimation and vice versa (Gupta et al., 1999). PBIAS is given as [%] by:

$$PBIAS = \frac{\sum_{i=1}^{n}\left(P^i_o - P^i_m\right) * 100}{\sum_{i=1}^{n}(P^i_o)}$$

where P^i_o is the observed and P^i_m is the modelled parameter at the time step i.

According to the models reviewed by Moriasi et al. (2007) a NSE of >0.80 and a PBIAS of >15 % appeared satisfactory for models with monthly time step. Thus these values were set as minimum calibration termination goal.

The calibration eventually achieved a best fit of observed and simulated dam storage for transmission losses from runoff of 32.6 % in the wet year (2002/03), 7.7 % in the dry year (2003/04) and 14.2 % in the average year (2004/05) and a dam seepage ratio of 9.8 % of stored volume (with a minimum storage threshold of 0.45 MCM stored water). **Fig. 3.27** shows the graphical representation of the calibration and validation results for the stored reservoir volumes between 2001 and 2009. For the calibration period a very good fit could be obtained with the mentioned differential transmission losses for the water years and the respective parameters were applied to the validation period with respect of the water year condition (wet, dry or average). The statistical evaluation criteria NSE and PBIAS for the calibration and validation period are given in **Table 3.13**. For the calibration period both criteria show a very good fit with a NSE >0.95 and a PBIAS <10 %. For the validation period the NSE still performs very well with values >0.90, whereas the years 2006-2009 give a somewhat poorer performance of PBIAS >10 %, but still are within the range of satisfactory model performance. According to the PBIAS criteria the simulations appear to have a general trend of slight underestimation, whereas the NSE values express a good overall fitting of simulation and observation trends.

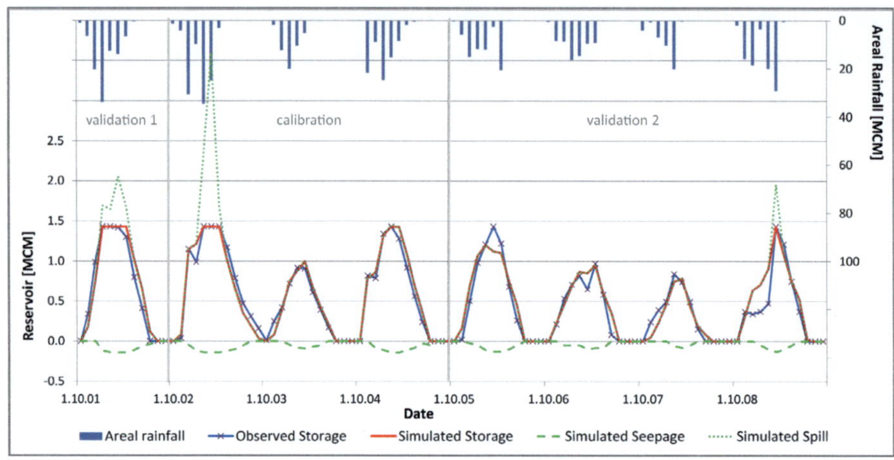

Fig. 3.27: Calibration and validation results of observed and simulated stored volumes in the Wadi Shueib Reservoir for the period from 2001-2009. The water years from 2001/02-2004/5 were used for calibrating transmission losses of surface runoff and the dam seepage ratio.

Table 3.13: Model performance indicators for the simulation of storage volumes in the Wadi Shueib Reservoir for the calibration and validation period 2001-2009.

Year		NSE [Dimensionless]	PBIAS [%]
2001/02	*validation*	0.96	-3.9
2002/03	*calibration*	0.97	5.6
2003/04	*calibration*	0.96	-0.5
2004/05	*calibration*	0.97	-8.2
2005/06	*validation*	0.93	-7.9
2006/07	*validation*	0.93	-10.2
2007/08	*validation*	0.92	10.8
2008/09	*validation*	0.90	19.1
2001-2009		**0.95**	**-1.9**

According to the validation test the model appears to provide a good prediction of the total annual received water in the Wadi Shueib Reservoir. Initial fill or final exhaust of the reservoir appears to be delayed by a monthly time step for the validation years 2001/02, 2006/07 and 2007/08. The peak storages appear well simulated for all years except 2005/06 where the simulation does not reach the observed storage level. Also the often occurring partial drainage of the reservoir after initial fill, as displayed in the observed storage, is generally neglected in the simulations. This is most probably due to a dam operation habit that is not logged properly in the dams discharge records. As dam operation is not the intended model use, this error was accepted. Overall the model displays satisfactory performance for the assessment of the monthly integrated catchment water balance. It is, however, obvious that the calibrated transmission loss parameter might aggregate various other estimation errors. In order to check the model for its validity in its intended use as tool for simulating effects of water allocation changes the respective parameters were subjected to a sensitivity and uncertainty analysis.

3.3.7.4. Model Uncertainty

An integrated model of the multiple water sector components as necessitated for the IWRM approach necessarily means a lot of assumptions and estimations. Beyond the evaluation of validity an important criterion for any model applicability is the assessment of the uncertainty and the sensitivity of model in- and outputs. As various authors have pointed out, even models of satisfactory calibration and validation can yield large uncertainties in their predictions due to non-uniqueness of parameter estimations (Beven, 2006; Kuczera & Mroczkowski, 1998; Vogel & Sankarasubramanian, 2003). It is obvious that such equifinality increases with the number of employed parameters and the model complexity. However, even rather modest lumped models with few parameters have been experienced to produce similar results for variable parameterisations (Gupta & Sorooshian, 1983). While inherent in any modelling approach, it is important to assess this uncertainty and provide it as information to decision processes based on the model predictions. Typically there are two parallel approaches to identify and understand the residual uncertainty in a model: sensitivity analysis and uncertainty analysis.

The objective of a sensitivity analysis is to explore and quantify the impact of errors in input data or parameter estimates on the model predictions and it is often done by systematically running simulations with alternative assumptions (Loucks et al., 2005). Model uncertainty is commonly assessed by describing the, often probabilistic, distribution of measured input data errors or parameter estimates and their effect on model output uncertainty (Loucks et al., 2005), as well as the identification and description of approximations and knowledge gaps.

Most monthly hydrological models employ only areal precipitation and potential evapotranspiration as fixed inputs and rely on a variable number of estimated parameters for their simulations (Xu & Singh, 1998). In comparison thereto the Wadi Shueib WEAP model, as a lumped water balance model dedicated to allocation simulation, was set up with a wider set of fixed inputs records and a likewise reduction of modelled physical relationships and required parameters. The reliance on recorded input data does, however, not imply any superior model quality, but rather transports the uncertainty range from the approximate nature of parameter estimation towards the domain of measurement uncertainties, which are in fact much more difficult to assess. Various aspects in relation to the uncertainty of the input data have already been discussed in Chap. 3.3.4 and for the pre-processing procedures in Chap. 3.3.5.

3.4.   Wadi Shueib Planning Scenarios

With the aim of testing the Wadi Shueib WEAP model as a practical IWRM decision support tool, it was employed exemplarily within a scenario planning exercise for the catchment area. In order to reflect a realistic framework, the simulated catchment management plan was aligned to the implementation approaches of the Jordanian national water strategy. An alternative business as usual strategy was simulated for comparison. With regard to the planning horizon of the Jordanian water strategy, which is currently the year 2022, the scenario horizon simulated in the Wadi Shueib WEAP model was set to 2025.

In order to define the definite objectives and implementation strategies of the Jordanian water strategy and to extract the relevant issues for the study area, a preliminary analysis of the published policy documents and action plans was undertaken. In a second step a set of indicators was established which should feature the ability to comprehensively evaluate the performance of the alternative strategies in terms of the stated objectives. Two alternative development lines of climate, population and water use were taken into account in order to address the uncertainties about the future water resources pressures. Eventually the water strategy was implemented into the Wadi Shueib WEAP model and the performance indicators were simulated and set in comparison to the business-as-usual strategy.

3.4.1.   Objectives and Plans of the Jordanian Water Strategy

With the ambitious new water strategy "Water for Life" the government of Jordan presents its vision of the goals to reach until 2022 in each of the major areas of the water sector. The strategy clearly follows the IWRM paradigm of sustainable management under social, economic and environmental considerations overarching water services and related sectors and adapting to the future challenges of population growth and economic development (RCW & MWI, 2009). The main topics addressed are the legislative and institutional set-up, the development and monitoring of available water resources, resource protection and sustainable management, efficiency of demand and supply management, quality standards, shared resources, public awareness, private sector participation, financing of water services under socioeconomic considerations, as well as major research and development issues.

The published strategy document is organized under the six topics of (i) Water Demand, (ii) Water Supply, (iii) Institutional Reform, (iv) Water for Irrigation, (v) Wastewater and (vi) Alternative Water Resources and is accompanied by a detailed Action Plan (MWI, 2009b). Within these chapters, however, it partially lacks a clear distinction between stated development goals, envisaged implementation approaches and explicitly proposed actions and also shows a lot of redundancies which hamper a systematic approach for IWRM planning and decision making. In 2011, a Conceptual Update to *"Water for Life"* was proposed within the USAID/Jordan Institutional Support and Strengthening Program (USAID/JISSP, 2011), which accounts for some of the mentioned issues as well as for recent changes in political, financial and physical conditions.

In order to define the distinct objectives for the planning scenarios in the Wadi Shueib case study, a distilled and reorganized version of the Jordan Water Strategy on the basis of the published strategy documents and personal interviews with stakeholders from the policy and decision making institutions was prepared. The schematic in **Fig. 3.28** depicts the main topics and their primary objectives and interactions as resultant from this analysis.

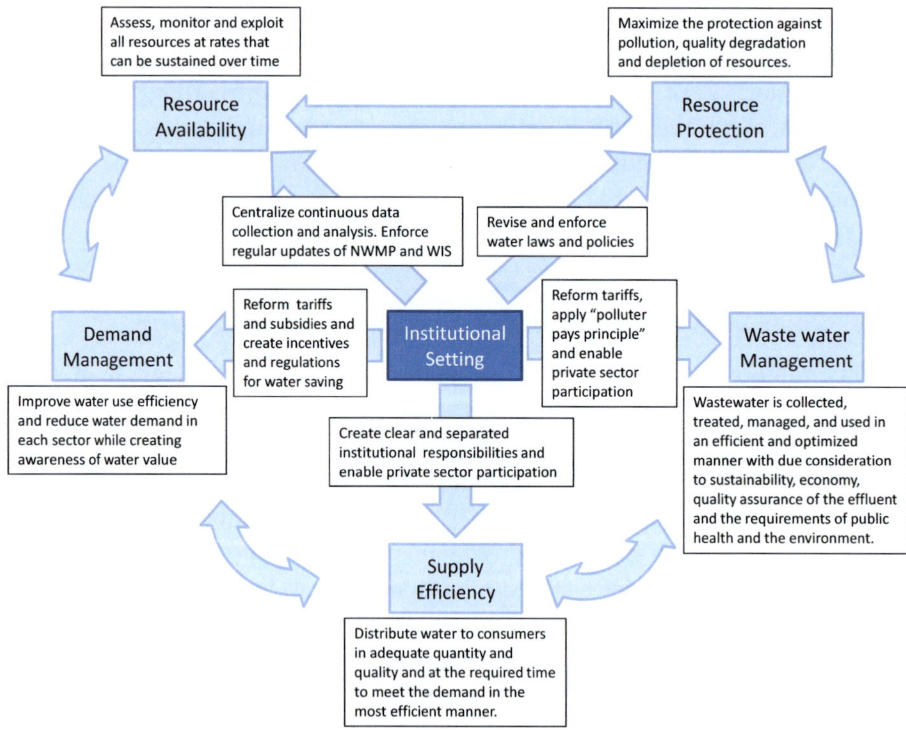

Fig. 3.28: Schematic of the topics addressed in Jordan's Water strategy and the respective primary objective statements.

A complete list of the identified objectives and implementation approaches according to the main topics can be found in Appendix A.

With regard to the situation in Wadi Shueib catchment not the complete range of objectives and implementations as stated in the National Water Strategy appears relevant. Thus, a subset of appropriate and applicable goals and measures was extracted which is listed in **Table 3.14** and the stated general implementation approaches were mapped on their possible effect within the Wadi Shueib water system. It has to be mentioned that the comprehensive set of implementation approaches as stated in the strategy could well provide thorough solutions for the water issues in the Wadi Shueib. However, the accompanying detailed action plan only foresees two actual projects in the study area (marked as bold in the Table 3.14). The presented list of objectives and actions henceforth built the basis for the further scenario development.

Table 3.14: Strategic objective and implementation approaches of the Jordanian Water Strategy and their possible impacts in the Wadi Shueib area.

Objective	Ind.	General implementation approach	Possible impacts in the Wadi Shueib
O.1: Increase volume of captured and treated wastewater	I.1a I.1b	Connect all households in urban areas served by treatment facility	**Increase the sewer connection ratio in As Salt**, Fuheis and Mahis
		Rehabilitate all sewerage pipes which are over 10 years old	Decrease the leakages from sewer pipes in As Salt, Fuheis and Mahis
		Install decentralized treatment to serve semi-urban and rural communities	Installation of DWWT units for Ira and Yarka
		Establish and enforce standards for the use of septic tanks in rural areas	Decrease the losses from septic tanks
O.2: Maximize Resources Availability	I.2	Use treated effluent in agriculture, industry, landscapes and aquifer recharge	The amount of treated waste water directly allocated to agriculture is increased
		Implement rainwater harvesting in new buildings and industry compounds	Fuheis Cement builds a rainwater harvesting pool
O.3: Secure constant drinking water supply	I.3	Expand production and supply networks to required capacities in all regions	Increase the treatment capacity of the Salt Springs Water microfiltration plant to meet future water needs
	I.4	Rehabilitate old and damaged supply sources and network components	**Reduction of the physical supply network losses**
O.4: Establish demand management		Conduct national programs for water education and conservation	The per capita domestic consumption does not rise above the targeted 120 l/c/d
O.5: Improve cost efficiency of supply and sanitation services	I.5	Improve meter reading and billing accuracy to reduce administrative losses	Increases of revenue per supplied m³ from the water treatment plants and the water imports
O.6: Protect water resources and environment	I.6	Establish protection zones for recharge areas of drinking water supply resources	Implementation of Spring Protection Zones for Azraq, Baqqouria, Hazzir and Shorea Springs
		Consider environmental quality and quantity demands in sensitive habitats when designing water allocation plans	Establish and monitor a minimum flow requirement for the perennial wadi stream.

3.4.2. IWRM Performance Indicators

In order to compare and evaluate the planning alternatives within the scenario simulations a set of suitable indicators has to be selected. Recommended frameworks for selection or development of sustainability and IWRM indicators, as well as proposed collections of indicators can be found in abundance in the literature (cf. Chap. 2.3.2).

Although valuable as starting point, it appears that the majority of general environmental indicator frameworks primarily concentrate on the development of simplified conceptual models and causal chains or networks (e.g. the PSR and DPSIR models; cf. Chap. 2.3.2). The purpose is hereby usually to narrow down very encompassing objectives, like "sustainable development" or "IWRM", to receive a structured set of categories and subjects of interest (e.g. Bossel, 1996). However, the ultimate selection of an appropriate distinctive measure is typically beyond the frameworks and thus left to each case studies own consideration. Consequently the number of proposed IWRM indicator sets comes close to the number of undertaken practical IWRM studies in total. At the same time related studies point out that the quality of the indicator selections relies primarily on how precise objectives and decision questions can be defined (e.g. Besleme & Mullin, 1997; Bossel, 1999; Niemeijer & de Groot, 2008). Vague or overly broad problem formulations (e.g. "loss in biodiversity") are of little use in selecting indicators (Segnestam, 2002).

Ergo, the mentioned indicator frameworks mostly apply themselves to identify and order critical factors from the wide spectrum of theoretically possible things to assess in complex systems. Readily available indicator sets on the other hand are either rather general or if specific usually hardly fit to the objectives of an explicit planning and decision situation.

In the case of the strategic scenario planning exercise for the Wadi Shueib, the objectives have, however, already been specifically defined from the Jordanian National Water Strategy as detailed IWRM policy guideline (cf. Chap 3.4.1). Likewise, an attempted application of the DPSIR framework as structural framework for the participatory indicator selection process, which involved project partners and stakeholders from the official institutions, led to some drawbacks. In accordance with the criticism formulated by Niemeijer and De Groot (2008), participants found it often difficult to follow the unidirectional structure of the approach. Although, the use of the DPSIR framework provided a simple interface for early communication between researches, stakeholders and decision makers, superior outcomes could be generated in discussions directly based on the conceptual WEAP models.

The actual selection of indicators was therefore organised in direct reference to the stated objectives and the catchment model and discussed during several sessions with project partners and stakeholders. Despite the well elaborated objectives, the selection was still not trivial and was guided by a range of quality criteria based on the related literature (Bossel, 1999; Niemeijer & de Groot, 2008; Segnestam, 2002) as well as constraints inherent in this studies framework:

Purpose: Naturally, the leading consideration has to be that of the actual purpose of the indicator, be it e.g. monitoring of current system states or trends, comparative developments on larger scales or specific performance of planned and undertaken actions in relation to goals and targets. The latter applies in the Wadi Shueib use case.

Goal relevance: As the ultimate goal of strategic scenario planning is to support the decision making process, it is of primary importance that the selected indicators can be directly used as decision criteria for the objectives addressed.

Decision maker relevance: The usual range of different perspectives from scientist over practitioners to policy makers holds the risk for indicators missing the decision relevant information.

Clarity: It is important that indicators are unambiguously defined and offer clarity in their interpretation. For the decision maker this means a solid scientific basis as well as transparency of its underlying assumptions and meaning. This often means to find a balance between a complexity that fully covers a system element and ease of understanding.

Computability/Measurability: Naturally an indicator has to practically assessable and quantifiable (at least normatively) with appropriate means. In the Wadi Shueib use case this basically meant that the indicators should provide the possibility for simulation with the available model capabilities.

Number of indicators: A very important criterion for decision making indicators that appears often neglected is the extent of proposed indicators sets. It is not unusual for indicator sets in the environmental assessment domain to comprise more than 50 or even above 100 recommended measures. For an extreme case see for example Bossel (1996) who compiled a list of 215 sustainable development indicators of which he stated at least 84 as necessary. There are perhaps two main reasons for this tendency: on the one hand scientists understand indicators as a possibility to communicate their system understanding and modelling capabilities to the public and to policy makers. On the other hand, it is the attempt to thoroughly cover all aspects of a natural system and to minimize uncertainties, which almost unavoidably leads to an ever growing number of suitable measures.

It is obvious, that in the case of real planning and decision situations an overly large set of indicators necessary to cover the intended objectives consequently means a large set of decision criteria, which then may rather hamper the decision process. This perhaps also contributes to the common IWRM critique of over-integration and being idealistic and non-achievable (e.g. (Biswas, 2004).

Non-Redundancy: Especially when intended as decision criteria it appears sensible, that the indicator set avoids redundancy in relation to the objectives in order to avoid a weight bias. Interrelations between indicators on the other are of no concern.

The indicator set eventually selected for this study has to be understood as a minimum set with the purpose to directly and efficiently assess the objectives of the water strategy. Following the full IWRM principles several additional indicators would need consideration.

3.4.2.1. I.1a/b: Municipal Waste Water Treatment/Recharge Ratio

According to the stated objective of increasing the volume of captured and treated waste water this indicator comprises the assumed volume of total waste water produced with the amount of municipal waste water treated in centralised and decentralised treatment facilities, also including the volumes pumped from the septic tanks and cesspits in urban and rural areas with

$$TR_{MWW} = 100 \times \left[\frac{WW_{TP} + WW_{DTP} + WW_{CPpumped}}{WW_{cap} \times Pop_{total}} \right]$$

where TR_{MWW} is the municipal waste water treatment ratio [%], $WW_{TP+DTP+CPpumped}$ are the inflow volumes of centralised and decentralised waste water treatment plants and the pumped sewage from cesspits [m^3] , WW_{cap} is the average volume of generated waste water per capita [m^3/capita/month], Pop_{total} is the total population in the catchment area [capita].

The volume of per capita generated waste water is here for estimated from the waste water treatment plant inflow BOD parameters (cf. Chap. 3.2.8).

In terms of health and resources protection it is furthermore necessary to assess the potential impact of discharged untreated waste water. This is not only related to the total discharged volume, but also to the dilution effect within the environment. As the majority of untreated waste water in the study area is assumed to infiltrate into the underground, a second indicator is proposed that sets the annual discharge of untreated waste water in relation to the estimated amount of natural groundwater recharge, given by

$$WW_{recharge} = 100 \times \left[\frac{WW_{cap} \times Pop_{total} - \left[WW_{TP} + WW_{DTP} + WW_{CPpumped} \right]}{I_{Prec}} \right]$$

where WW$_{recharge}$ is the ratio of untreated waste water to natural groundwater recharge [%], I$_{Prec}$ is the annual volume of groundwater recharge as estimated from the water balance modelling [m³].

3.4.2.2. *I.2: Available Renewable Water for Internal Use*

The annual per capita total renewable water resources volume is frequently used (in slightly different forms) as a direct measure for the resource pressure within a given area (Falkenmark, 1986; Gleick, 2003; UNEP/GRID-Arendal, 2008). Most of these indicators are assessed on a national or basin-wide scale and are often used for inter-basin comparisons regarding actual and future water stress and potential for conflict (Yoffe et al., 2003). Perhaps most often cited in this respect is Falkenmark's Water Stress Index (Falkenmark, 1989) which divides a countries total available water resources in the form of net precipitation and water imports by its population and relates the resultant per capita availability to an estimated gross per capita water requirement (including household, agricultural, industrial and environmental needs) of 1,700 m³ per year. Countries with water resources below this threshold are understood as water stressed, water scarce (<1,000 m³/capita/year) or absolute water scarce (<500 m³/capita/year) (cf. Chap. 1.1). One of the main criticisms on the Water Stress Index is that it does not take the true availability for use, with respect to infrastructure, of the water resources into account, nor does it account for a countries ability to adapt and to find alternative water sources (Rijsberman, 2006). These, however, are crucial points for water management planning studies.

Accounting only readily developed (available for use) renewable resources gives a better image of the true resources development in an area which is desirable for assessing the strategic objective of maximising the availability. Thus, for this study the available resources were considered as the total sum of water volumes produced from renewable sources, including surface runoff, groundwater, reclaimed water and harvested rainwater. With regard to the sources and consequently the quality types the indicator comprises sub-indicators for groundwater, stored runoff and reclaimed water and is applied in the Wadi Shueib case study by

$$ARW_{intern} = \frac{1}{Pop_{total}} \sum_{QT=1}^{3} P_{QT} + I_{QT} - E_{QT}$$

where ARW$_{intern}$ is the per capita available internal renewable water [m³/capita/year], Pop$_{total}$ is the total population in the catchment area [capita], QT is the quality type (groundwater, surface runoff, reclaimed water), P$_{QT}$ is the total annual production of the respective quality type [m³], I_{QT} is the total annual import [m³] and E_{QT} is the respective total annual export [m³].

3.4.2.3.   I.3: Water Supply Shortage Index

To assess the unmet demand of water supply it is important to consider quantity as well as frequency of supply shortages. The water supply shortage index was originally developed for dam operations to assess annual shortages over multiyear periods (Srdjevic, 1987). It is also convenient for to assess monthly shortages of any other supply framework as it acts as an indicator of both frequency and quantity of shortages. It is given by

$$WSS_{DS} = \frac{100}{n} \sum_{i=1}^{n} \left(\frac{S_i}{D_i}\right)^2$$

where WSS_{DS} is the water supply shortage index of a certain demand site or sector [dimensionless], n is the number of time steps in the assessed period [e.g. months], S_i is the total quantity of unmet demand in the time step i [m³] and D_i is the total demand in the time step i [m³].

3.4.2.4.   I.4: Water Supply Requirement

In the light of Jordan's current situation, where water demand exceeds available renewable resources, the establishment of adequate demand management measures appears vital for any future development. The stated objective of demand management might be misleading, as the essential goal is actually to reduce the amount of required supply. This may involve the reduction of "actual water demand" through various approaches (e.g. enhancing water awareness, water saving incentives and improved efficiency of domestic, irrigation and production technologies), as well as the decrease of required supply by internal reuse, water harvesting and the reduction of distribution losses and avoidable water wastages (Twort et al., 2000).

In scarce environments, the actual water consumption is of course mostly dependant on the offered supply. Thus, in order to assess a supply requirement it is necessary to preliminary set a target volume for the actual use, in other words the water demand, e.g. the anticipated volume of per capita available water for domestic consumption or the volume necessary per agricultural area or production unit. Water efficiency enhancements or public water awareness programmes aim at directly reducing the water demand, whereas reuse and loss reduction efforts attempt to reduce required supply. Thus, the water supply requirement (WSR) is given by

$$WSR = \frac{(D_{au} \times A_{au} \times (1 - W_{reuse}) - W_{alternative})}{(1 - Loss_{supply})}$$

where WSR is the water supply requirement for a certain demand site or sector [m³/time step], D_{au} is the actual demand per activity unit (e.g. per capita, production unit or agricultural area) [e.g. m³/capita/time step, m³/ha/time step or m³/production unit], A_{au} is the total activity [e.g. capita, ha or production/time step], W_{reuse} is the process water reuse or recycling ratio [%], $W_{alternativr}$ is the volume of water from alternative sources (e.g. from water harvesting) [m³] and $Loss_{supply}$ is the ratio of internal or supply system water losses [%].

3.4.2.5.   I.5: Full Water Service Cost

A basic economic measure to assess the cost effectiveness (also: *least cost analysis*) of alternative investment options in the water sector is the unit cost of supply and sanitation. It can be appropriately applied in cases with pre-set objectives, as e.g. with the anticipated water supply and sanitation goals in strategy planning. As "Full Service Cost" per supplied cubic meter at the demand site the unit cost has to include all capital, operation and maintenance costs for production, distribution, return flow collection and treatment. By accounting the unit cost at the demand side, distribution and return flow losses are merged into the cost figures. Capital costs are typically annualized over the expected life span of an investment and combined with yearly operation and maintenance costs to account for the time value of money (Asian Development Bank, 1999; Fane et al., 2003). The annuity capital investments is given by

$$ACI = C_{PV} \times \frac{r}{1 - (1 + r)^{-n}}$$

where ACI is the annualized capital investment [JD], C_{PV} is the capital investment as present value [JD], r is the nominal annual discount rate [%] and n is the expected lifespan of the investment [years].

The unit costs for supply and sanitation of a water commodity (e.g. drinking water) is then given by

$$FSC_{UC} = \frac{1}{Q_D} \times \sum_{i \in I} [ACI_i + O\&M_{fix_i} + O_{var_i} \times Q_{T/P_i}]$$

where FSC_{UC} is full service cost as unit cost [JD/m³], Q_D is the volume delivered at the demand side [m³], I is the complete set of supply and sanitation infrastructures used for delivering the commodity, ACI is the annualized capital investment cost for an infrastructure element [JD], O&M$_{fix}$ is the fixed operation and maintenance cost [JD], O$_{var}$ are the variable operation costs [JD/m^3] and Q$_{T/P}$ is the production, flow or treatment volume at the respective infrastructure element [m^3]. The full service costs can easily be split into sub-indicators for water supply and sanitation costs by accounting only the particular infrastructures.

It is important to note that the proposed cost indicator can only be understood as a relative metric for reflecting the influence of investment decisions on the direct water service costs within the boundaries of an autarkic system. In order to achieve a full economic cost assessment as well as an understanding of the economic value of the water resources it would be furthermore necessary to include the opportunity costs of alternative supply options as well as eventual economic and environmental externalities (Rogers et al., 1998).

3.4.2.6.   I.6: Environmental Water Stress (with return flows)

Accounting for the environmental response to short or long-term water scarcity or water stress situations is difficult due to the multiplicity of involved factors as well as the natural adaption ability. An accepted concept in ecology is that the preservation of aquatic ecosystems should be considered according to their natural flow regime variability (Hughes & Hannart, 2003). Based on this assumption, Smakhtin et al. (2004) proposed a relatively pragmatic approach for defining the environmental water stress of a river basin by setting the actual runoff in relation to the natural high and low flow volumes. For that purpose they defined the environmental water requirement (EWR) as sum of the Q90-flow (monthly flow that is in average exceeded 90 % of the time during a year) and a high flow volume which is estimated to 5 – 20 % of the natural mean annual runoff depending on the Q90-flow fraction (Smakhtin et al., 2004). With regard to the EWR, the environmental water stress indicator (WIS) is given by

$$WSI = \frac{total\ withdrawals}{MAR_n - EWR}$$

where WSI is the environmental water stress indicator of the catchment [dimensionless], MAR$_n$ is the (natural) mean annual runoff [m^3] and EWR is the environmental water requirement [m^3].

The categories of the resulting WSI are proposed by (Smakhtin et al., 2004) with

$$WSI = \begin{cases} > 1 & \rightarrow \quad \textit{Overexploited and environmentally water scarce} \\ 0.6 < WSI > 1 & \rightarrow \quad \textit{Environmentally water stressed} \\ 0.6 < & \rightarrow \quad \textit{no water stress} \end{cases}$$

In the presented form the WSI accounts for total withdrawals but does neglect return flows, which may be satisfactory for systems with predominantly agricultural consumption. In the Wadi Shueib catchment the return flows of treated waste water deliver a significant proportion of the dry weather flow. Also, the primary withdrawals (from springs) and return flows (from treatment plants) occur in relatively close range in the catchment area. Thus, to consider the return flow volumes for this study the WSI was adjusted to be used in the form

$$WSI_{RF} = \frac{MAR_n - AR_o}{MAR_n - EWR}$$

where WSI_{RF} is the adjusted WSI to consider return flows [dimensionless], MAR_n is the natural mean annual runoff [m³], AR_o is the actual observed (or simulated) runoff [m³] and EWR is the environmental water requirement [m³].

The WSI or WSI_{RF} offers a relatively easy-to-assess environmental indicator which was convenient for this study but some important shortcomings have to be kept in mind. The WSI is based on annual mean runoff and does not consider amount and frequency of monthly or seasonal water shortages. Furthermore, it is exclusively an indicator for quantitative shortages but does not consider environmental water quality requirements. Smakhtin et al. (2004) also mention that the WSI does not specify a desired conservation status in which an ecosystem needs to be maintained. In order to address these issues it would be necessary to employ additional environmental indicators that should best be adapted to the actual local ecosystem conditions.

3.4.3. Alternative Development Scenarios

With regard to the ambitious goals set by Jordan's national water strategy it is obvious that their achievement involves numerous uncertainties. Key factors hereby include external driving forces such as the natural annual variability of available water resources due to climatic conditions as well as the demographic and socio-economic development within the coming decades that might considerably change the demand patterns in the different sectors. The main internal driver in the process will most likely be the eventual political and public determination, effort and ability to realize all necessary steps mentioned in the implementation approaches. These factors can hardly be directly influenced by the decision makers in the planning process, yet their imminent uncertainty has still to be taken into account in a sound and sustainable planning approach.

To account for these uncertainties in the planning process, the identified key driving forces were used to basically define two thinkable development outlines for both the external as well as the internal drivers. The basic development narratives are as follows:

The **High Resource Pressure (HRP)** development is characterized by a relatively dry period over the coming 15 years with an above average frequency of dry water years. Furthermore, the population growth follows the recent trends and remains at a very high level. No significant decrease of gross domestic product (GDP) in agriculture is observed, thus the agricultural production stays on the current level. On the other hand the GDP in the industrial and the service sector is increasing, resulting in a change of lifestyle and a fast increasing per capita water demand in the municipal and the industrial sector.

The **Low Resource Pressure (LRP)** development is characterized by a climatic period in the range of the long term average observed in the area. The demographic transition takes a faster pace, resulting in a decreasing population growth towards medium rates. Agricultural production and water demand is decreasing within the area and a rise of demand in other sectors is observed, which is, however, less dramatic in comparison to HRP.

In the **Business as Usual (BAU)** development the implementation of the national water strategy in the Wadi Shueib area is realized within the range of the already active projects and their current performance as reported in the annual MWI reports (MWI, 2010). This comprises the reduction of physical and administrative supply network losses and a sewer rehabilitation and connection program in As-Salt (the projects marked as bold in the third column of Table 3.14). In this development line it is assumed that the implementation of further strategic objectives of the water strategy appears to be either not feasible until 2025 or is hampered by slow political decision making.

The **Full Implementation (FI)** development line assumes that all obstacles are overcome and the full range of stated implementation approaches is realised until 2025 in the Wadi Shueib area (the complete third column of Table 3.14).

The development outlines were eventually grouped in a set of four planning scenarios as depicted in **Fig. 3.29**. The combinations, even though reasonable, are of course arbitrary and the scenarios have to be understood as not necessarily mutually exclusive and exhaustive set of future states. It is often emphasized that scenarios have to be perceived as equally valid (although not necessarily equally likely to occur) and that therefore a truly robust strategy must allow the objectives to be achieved under all of them (e.g. IPCC, 2007; van der Heijden, 1996).

The fifth "reference" scenario is used for comparison and is merely based on the projection of the current trends of the external drivers into the future and assumes that no strategy implementation at all takes place.

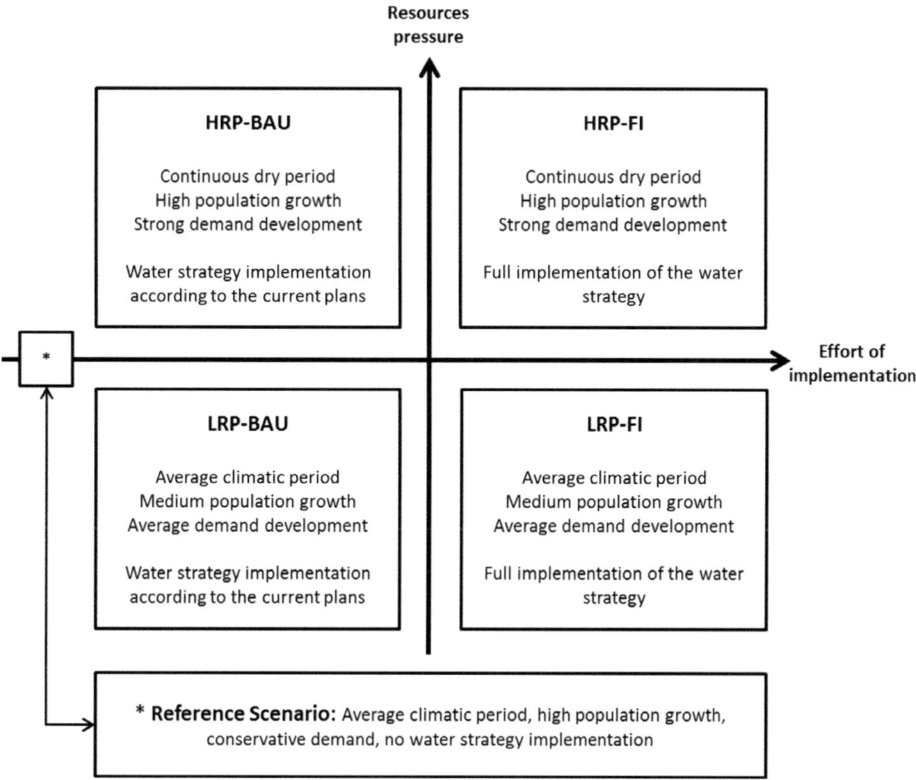

Fig. 3.29: Alternative development scenarios for the case study with a planning horizon of 2025.The scenarios are characterised by the development of the external and internal drivers towards a future of high or low resources pressure (HRP or LRP) and towards a business as usual or a full implementation of the water strategy objectives (BAU or FI).

3.4.4. Driving Forces

In order to translate the narratives of the alternative development scenarios into model assumptions, the reasonable development range of the external and internal driving forces was assessed on the basis of the current trends as observed in the records and the published literature.

3.4.4.1. Precipitation

According to the large scale studies on regional climate change undertaken by the IPCC (Nakicenovic et al., 2000), there is no clear projected trend in annual precipitation volumes in the Middle East. In accordance thereto, a comprehensive national study that compared the outputs of three Global Change Models

131

(CSIROMK3[5]; ECHAM5OM[6]; HADGEM1[7]) on the area of Jordan found that the respective precipitation scenarios are highly variable without a significant trend in volume or distribution (MWI, 2009). Several localized studies have found that the majority of rainfall stations in Jordan show a general decreasing tendency in annual precipitation volumes (Hamdi et al., 2009; MWI, 2009).

Besides the general trend of average parameter values, a robust planning scenario principally requires recognition of interannual variability and occurrences of extreme values. A pragmatic approach was chosen for the precipitation projections in the Wadi Shueib planning scenarios by adjusting the projections on basis of the historical records of annual precipitation volumes. As depicted in **Fig. 3.30** over the past 40 years the occurrence of average, dry and wet water year conditions has been relatively balanced with 51 % average, 26 % wet and 23 % dry water years and an average areal rainfall of 392 mm.

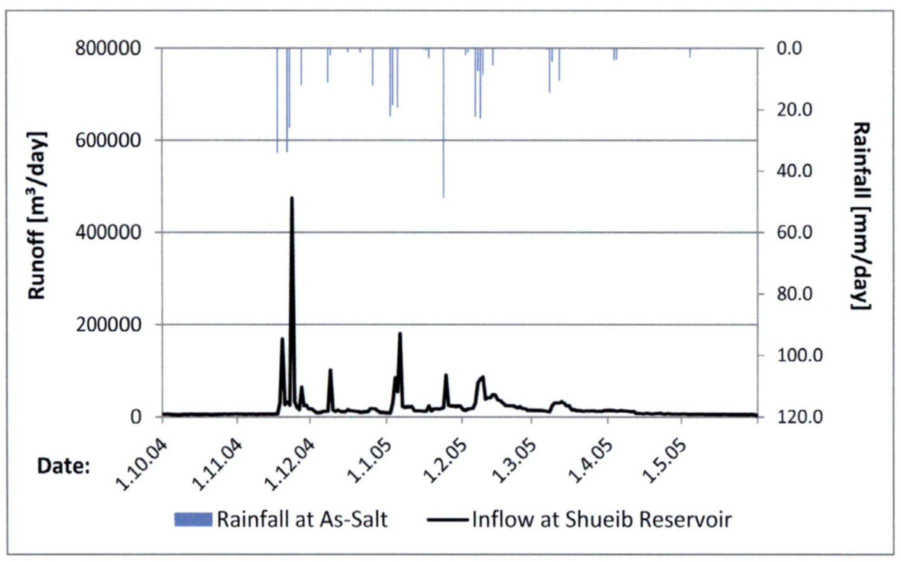

Fig. 3.30: Time series of interpolated annual areal rainfall volumes from 1970 to 2009 (based on station-records from the MWI and NN interpolation cf. Chap. 3.3.5.1).

---

[5] CSIROMK3: Commonwealth Scientific and Industrial Research Organization (CSIRO) Model, Australia.

[6] ECHAM5OM: The 5th generation of the ECHAM general circulation model, Max Planck Institute for Meteorology, Germany.

[7] HADGEM1: Hadley Center Global Climate Model, UK.

132

The last 15 years (1994-2009) can be considered as rather dry period with 33 % dry and 20 % wet year conditions and an annual average of 328 mm. For the "dry future" scenarios (HRP), it was assumed that the observed dry period continues for the next 15 years (2010-2025). Thus, in order to create the precipitation time series for this scenario, the annual rainfall volumes from the 1994-2009 period were randomly assigned to the years 2010-2025. In order to create an annually comparative time series for the "average precipitation future" scenario (LRP), the dry scenario time series was then simply shifted about 25 % towards higher rainfall amounts, resulting again in a balanced distribution of water year conditions. The resultant precipitation projections are depicted in **Fig. 3.31:** . The monthly distribution of rainfall amounts in the projections was kept unaltered as in the historical records.

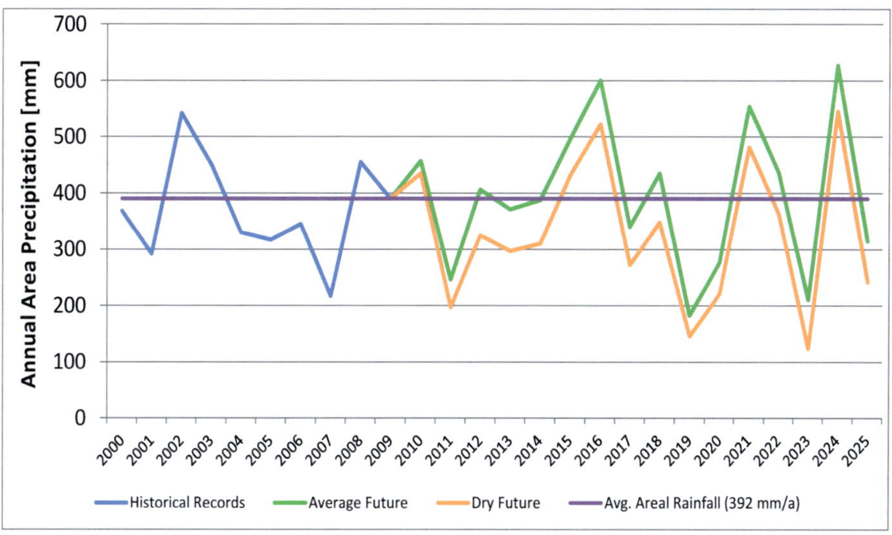

Fig. 3.31: Projected precipitation time series for the WEAP scenarios from 2010-2025.

3.4.4.2.   Temperature and Evapotranspiration

In comparison to the annual precipitation volumes and distribution the regional and local studies show more unity with regard to temperature trends and projections. The latest national climate change study comes to the conclusion, that the estimated changes in annual average temperature according to the large scale GCM (cf. Chap. 3.4.4.1) varies between +1.0 to +1.3 °C until the year 2050, whereas summer warming is expected more substantial than winter warming (MWI, 2009). In their comprehensive analysis of historical station records in Jordan Hamdi et al. (2009) found that annual maximum air temperature records did not show clear trends, but annual minimum temperatures have apparently increased while the annual temperature range has decreased since the 1970s.

In order to acknowledge the estimated temperature changes in the Wadi Shueib WEAP scenarios, the HRP scenarios were projected with an increase of 0.5 °C and 1 °C of minimum winter temperatures and minimum summer temperatures respectively in comparison to the areas long term averages. The reference evapotranspiration time series was then calculated as described previously in this study (cf. Chap. 3.3.5.2) but with the monthly averages of the historical humidity and wind speed records. The resultant reference evapotranspiration values for the dry year scenario ranged about 1-3 mm per month above the long term average values.

3.4.4.3.   *Population*

Jordan has experienced a tremendous population growth during the last decades and the Wadi Shueib area actually even lies above the country average (cf. Chap. 3.2.7). Even though the average population growth rate has declined in the periods between the last two censuses in Jordan (4.4 % between 1979 and 1994 and 2.5 % between 1994 and 2004), with regard to the current age structure, with a very high proportion of children to the total population in the Jordanian society, a very high growth trend is likely to continue (DoS, 2004). As future growth rates are not only dependent on the current demographic situation but also on the choices and lifestyles of the future population (especially of the fertile age range), there are usually different population growth scenarios to be considered. Population growth scenarios for the complete Jordanian society are given by the Higher Population Council of Jordan (HPC, 2009), the Department of Statistics of Jordan (DOS, 2010) as well as by the Population Division of the Department of Economic and Social Affairs of the United Nations (UN, 2010b). **Table 3.15** shows a comparison of the projections according to the different sources and their scenarios.

As the HPC population projections are currently the official policy planning scenarios for the Jordanian Government they were also adopted for the Wadi Shueib study.

The three HPC scenarios are distinctive in their assumption of at which point of time the demographic transition in the Jordanian society will result in a stable population replacement rate of 2.1 children per woman (2030 for Scenario II; 2040 for Scenario III). The HPC-Scenario I assumes a continuation of the current trend and a stable fertility rate of 3.6 children per woman in the forecasted period.

Table 3.15: Jordanian population growth rate projections from the last census in 2004 until 2030 according to different sources and scenarios. The HPC scenarios are the currently official planning scenarios for the Jordanian Government.

Source	Scenario	2005-2010	2010-2015	2015-2020	2020-2025	2025-2030
DOS (2009)	DOS Projection	2.34%	2.50%	2.08%	1.78%	1.54%
HPC (2009)	HPC-Scenario I	2.53%	2.55%	2.49%	2.34%	2.20%
	HPC-Scenario II	2.39%	2.15%	1.77%	1.56%	1.36%
	HPC-Scenario III	2.39%	2.15%	1.78%	1.62%	1.47%
UN-DESA (2010)	Constant fertility rate	2.94%	2.16%	2.08%	2.01%	1.93%
	High variant	2.94%	2.07%	1.90%	1.76%	1.58%
	Medium variant	2.94%	1.88%	1.61%	1.42%	1.25%
	Low variant	2.94%	1.69%	1.30%	1.04%	0.89%

For the Wadi Shueib scenarios the population figures of the municipal and rural population in the catchment area according to the last census (DoS, 2004) were projected until 2025 with the HPC-I rates as high population growth for the HPR-scenarios and the HPC-III rates as medium population growth for the LPR scenarios. The resultant population projections are depicted in **Fig. 3.32**. In the high growth scenario the population in the catchment area will reach 184,500 capita in 2025 which is an increase of about 70 % from 108,385 in 2004. For the medium growth scenario the total increase is about 56 % to reach 168,690 in 2025. This shows that even the official low or medium growth projections result in a high population pressure.

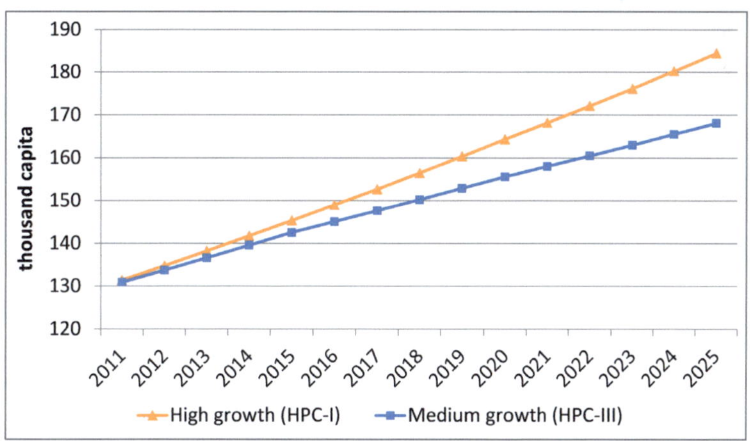

Fig. 3.32: Population scenario projections for the Wadi Shueib area 2010-2025. The growth rates are according to the HPC-Scenarios I and III.

3.4.4.4. *Water Demand*

In conjunction with its demographic and economic development Jordan has experienced an enormously increasing water demand during the last decades, especially in the municipal and agricultural sector. As shown in **Table 3.16** it is estimated that already today water demand exceeds total available resources (including waste water reuse and desalination) by more than 325 MCM per year and the gap is expected to continuously grow throughout the coming decades. Main water user still is the irrigated agriculture which has been politically supported and strongly subsidized during the second half of the last century. In the current national development plans, the municipal, industrial and tourism sectors have received higher priority, whereas agricultural supply ought to be stabilized at its current level (RCW & MWI, 2009). Despite the existing deficits and the dire future projections, the official policy still keeps a considerable focus on a supply side management strategy with the ambitious objective to not only secure the present per capita availability in the light of a growing population, but to even almost double the per capita available municipal water from the current level about 85 l/c/d towards a share of up to 150 l/c/d.

Besides a comprehensive re-allocation of freshwater resources from agricultural towards municipal uses, a considerable reduction of the abundant losses and the thorough development of all possible alternative resources, these plans are strongly dependent on the realization of mega-projects like the Red Sea-Dead Canal, whose implementation is currently by no means certain (Bdour, 2012).

Table 3.16: Historical and expected future development of sectoral demand and available resources in Jordan (table adapted from Haddadin et al, data sources WAJ).

	1990	2000	2010	2020	2040
			MCM/a		
Municipal	240	340	477	670	1263
Industrial	43	78	110	130	170
Agricultural	692	659	796	802	802
Total demand (incl. losses)	975	1077	1383	1602	2235
Total available resources	754	892	1054	1152	1549
Water deficit	221	185	329	451	687

Municipal Demand

Municipal consumption is expected to continue being the major water user in the Wadi Shueib area. According to the water supply and waste water balancing conducted in this study (cf. Chap. 3.3.5.5) the average municipal water consumption from network supply in the Wadi Shueib has been in the range of

85-95 l/c/d (excluding losses) in recent years. According to the official plans the municipal demand is targeted to rise to 150 l/c/d in urban areas and 132 l/c/d in rural areas by 2022 in order to meet the norms for a healthy population and rising living standards (MWI, 2004). It is, however, difficult to speak of water demand, as in a scarce environment consumption is of course more closely related to the actual supply. The last decades have seen a rather slow increase or during dry periods even a decline in per capita consumption due to the strict rationing of the water supply. Nonetheless, with regard to the intensified efforts of loss reduction in the supply systems it is assumed that the per capita water demand and actual consumption rises during the scenario period as depicted in **Fig. 3.33**. In order to account for the optimistic projection the demand in the HRP-scenarios reaches the officially anticipated 150 l/c/d in 2025, whereas the LRP scenarios assume a slower but nonetheless steady increase according to the overall trend of the previous decade which eventually reaches 117 l/c/d in 2025.

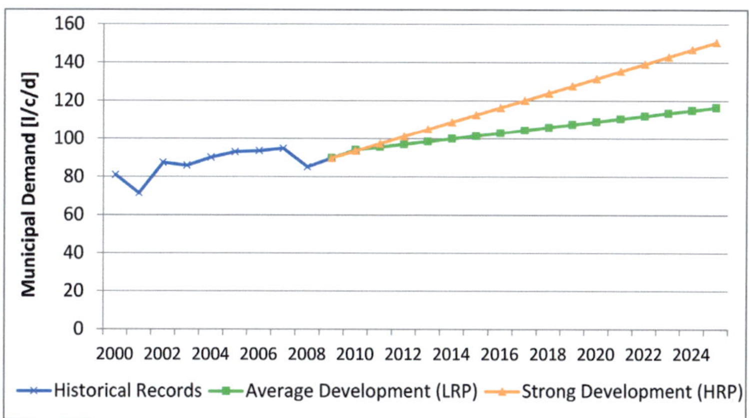

Fig. 3.33: Per capita municipal demand development in the Low Resources Pressure (LRP) and High Resources Pressure (HRP) scenarios.

Touristic Demand

Although the Wadi Shueib is favored recreation area for Jordanian families as well as a visiting spot for foreign tourists, there are no major touristic development plans for the area. Thus, the touristic demand is assumed to be covered by the municipal demand figures.

Industrial Demand

In the Wadi Shueib WEAP-model industrial demand is subsumed under the Lafarge-Jordan Cement factory node as major industrial actor in the area.

Due to technical problems in the production lines the factory has not been running on full capacity during the last ten years and the prevailing financial crisis led a reduction to less than half capacity in the year 2010 (Lafarge Cement Jordan, 2010). Furthermore, a new competitor is scheduled to enter the Jordanian cement market in 2013. On the other hand, the company made various major investments in recent years and with the international Lafarge company the Fuheis plant has a stable background. Thus, it appears unlikely that the location could be completely shut down. Nonetheless, according to the official strategic statement of Lafarge-Jordan Cement the priority in the near future is to reduce costs "*as a matter of survival*" and to secure the market position, rather than any expansion investment plans (Lafarge Cement Jordan, 2010). Regarding these considerations the industrial productivity is assumed to regenerate within the planning horizon to either reach the throttled levels recorded between 2005 and 2008 (LRP scenarios) or to the plants current full production capacity (HRP scenarios). Furthermore it is assumed, that due to the lack of major investments the current water demand of 340 litres per produced ton of cement remains constant (**Fig. 3.34**).

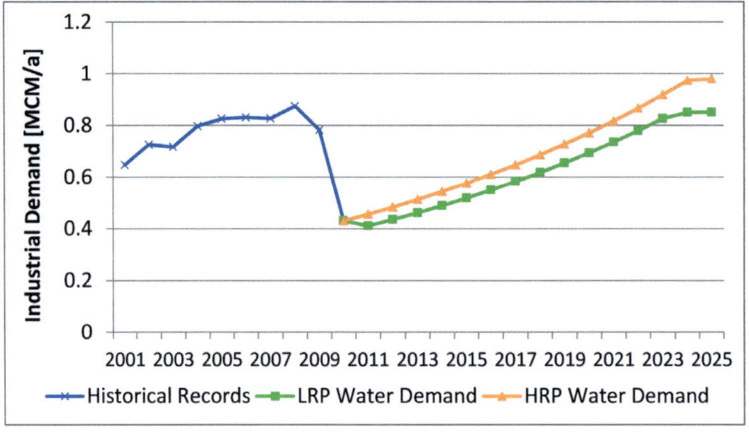

Fig. 3.34: Industrial demand development in the Low Resources Pressure (LRP) and High Resources Pressure (HRP) scenarios.

Agricultural Demand

The majority of agricultural activity in the Wadi Shueib area is rainfed olive trees and pastures, but some local landowners irrigate their plots with water from the smaller springs or the stream flow from the Wadi channel (cf. Chap.3.2.7). Furthermore, the waste water treatment plants of As-Salt and Fuheis have contracted some farmers in the vicinity of the plants to the use of a part of the treated effluent for irrigation. In respect to the given situation and the low priority of agricultural development in the official water agenda, it is assumed that the agricultural activity is not extended within the scenario planning horizon. On the other hand it is also not expected to experience a strong decline in the agricultural activity within the catchment, as most plots appear to be run as secondary activity by their owners and water supply is not controlled and rationed by the network supply system. Thus, the HRP scenario assumes no changes in the irrigated land area and the LRP scenario only assumes a slight decline of irrigation activity by 0.5 % annually. In a study of climate change impacts on the Jordan Valley region Menzel et al. (2007) simulated a decrease of 18 % of annual precipitation volumes to result in an increase of 22 % in irrigation demand. Even though these findings are difficult to transfer directly to the study area, it was attempted to incorporate this factor by assuming the irrigation demand to increase by 15 % until 2025, due to the assumption of prevailing dry conditions in the HRP scenarios.

3.4.4.5. *Internal Driving Forces*

As internal driving force this study basically understands the achieved grade of implementation of the water strategy. Although during the last years the responsible instituions have shown great will and ability for reforming the water sector it is still often critized that the sector appears too slow in response and that supply-side management strategies still dominate where demand management would be more rational and effective (Zeitoun et al., 2012). Reasons might be seen in the too large institutions, the lack of adequatly trained personell and the problematic of brain drain in Jordan, but also political and public opposition to some of the more uncomfortable changes in relation to traditional water rights.

Consequently, as described in the scenario narratives the BAU scenarios assume that the currently active projects are continued and finished until 2025 but new initiatives cannot be successfully undertaken within the study area. According to the annual water sector report (MWI, 2010) the achieved reduction of physical supply network amounted 0.8 % for 2009, whereas the objectives of reducing non-revenue water to 25 % by 2025 would need a target of annually 1.2 % loss-reduction. The sewer connection and rehabilitaiton project in As-Salt reported 103 new house connections and about 5,880 meters of implemented sewer length at a cost of approximatly 394 JD per meter. Assuming a total population of 91,000 capita in 2010 (cf. Chap. 3.4.4.3), an average of 6 persons per household (DoS, 2004) and the estimated connection degree of 69 % (MWI, 2010), this accounts for an increase of 0.9 %. In order to achieve the goal of full urban connectivity it would need an annual increase of 1.4 %.

For the FI scenarios it was assumed that the targeted values of physical loss reduction and sewer connectivity are achieved until 2025 in As-Salt and comparable projects are undertaken in the municipalities of Fuheis and Mahis. Furthermore, the smaller villages are served by decentralized waste water treatment facilities starting from 2015 with yearly increasing capacities. Details on possible implementation scenarios for decentralized waste water treatment in the study area were developed in a working group within the SMART research project (*see* (Müller et al., 2011)). The remaining cesspits are rehabilitated at a rate of 5 % per year. Further implemented projects include a rainwater pond at Fuheis Cement with a capacity of 15,000 m³ reducing the factory's network supply demand, a successful water awareness and demand management campaign that reduces municipal water consumption by 7 %, and the enforcement of the spring protection zones as proposed by BGR and MWI (2010) which results in the possibility to constantly exploit the full springs discharge for drinking water production.

In **Table 3.17** the assumptions on the development of the internal driving forces for the Reference, the Buisiness-as-Usual and the Full implementation scenarios are summarized.

Table 3.17: Assumptions on the development of the model parameters due to the different implementation scenarios.

	unit	2009	2025 Reference	BAU	FI
Supply and Consumption:					
Spring water use ratio*	%	46.5	46.5	60.0	90.0
Water harvesting**	MCM/a	0.0	0.0	0.0	~0.1
Administrative losses***	%	28.6	28.6	20.0	12.5
Physical network losses***	%	28.6	28.6	17.5	12.5
Municipal internal reuse	%	0.0	0.0	0.0	5.0
Sanitation:					
Waste water treatment coverage***	%	62.5	62.5	73.9	93.9
Leakage from sewer canals	%	5.0	5.0	3.0	0.5
Leakage from septic tanks	%	58.0	58.0	58.0	16.0
Treated waste water irrigation****	MCM/a	0.15	0.15	0.21	1.02

Sources: (Lafarge Cement Jordan, 2010; MWI, 2010; RCW & MWI, 2009)

*water produced against total discharge from Ain Azraq, Baqqouria, Hazzir and Shorea

**approximate amount provided by a rainwater harvesting pond at Fuheis Cement

*** in the BAU scenario, loss reduction and waste water treatment projects are realized in As-Salt only

**** direct application within the catchment

The necessary capital investments were inferred from literature values (mostly feasibility studies and project reports), either when available from within the study area or otherwise from comparable projects in Jordan.

Direct and detailed recommendations on investments for pumping and reservoir equipment of the main Wadi Shueib Springs and the Yazidiyya well field were available from an energy efficiency and performance study conducted within a German-Jordanian cooperation project on improved energy efficiency (GTZ et al., 2009). The assessment concluded that with an investment of about 2.2 Million JD into new pumping equipment and storage facilities the energy efficiency of the water production can be improved on average by about 20 %, which would payoff within 11 years.

Municipal Non-Revenue Water (NRW) is a major cost factor for the water supply and the need for investments into the reduction of physical as well as administrative losses is perceived as essential, both in economic terms as well as in the light of resources scarcity. However, despite considerable efforts and some successful pilot projects, Jordan has so far not been able to reach its goals in the reduction of NRW (Sommaripa, 2011). A successful reduction of NRW does not necessarily follow a straightforward technical investment strategy but rather requires an efficient composition of three elements: repair and extension of supply networks, improved operation and maintenance performance and improved efficiency of the public water utilities for metering and billing their subscribers. In Amman for example the public water company invested over 250 Million JD into network restructuring and rehabilitation to reduce the NRW from above 50 % to about 40 % (Sommaripa, 2011). On a smaller scale, a pilot study in Ayn al Basha (Amman suburb) demonstrated an approach on effective leak detection, repair and maintenance management that reduced water losses by about 20 % with a project budget of 0.35 Million JD (Kugler, 2010). In terms of reducing administrative losses, a GTZ and WAJ pilot project in Madaba showed that an outsourcing of subscriber management within a private sector partnership led to a direct payoff for the water company (Rothenberger, 2009). Initiatives to reduce NWR in the study area have, however, until now been marginal. Thus, the necessary investments in this segment for the FI scenario were inferred as a combination of the three above cited projects while their reported costs were scaled to the constraints of the study area.

Several projects have been addressing issues of demand management in Jordan, especially in Amman and Aqaba. One comprehensive approach is taken by the USAID-funded project for Instituting Water Demand Management in Jordan (IDARA) which also conducted a plumbing retrofit program in several urban and rural areas to reduce household water demand and internal losses. From the outcome of these retrofit programs it is estimated that with an investment of about 200 JD per household the domestic demand can be reduced by 11 % through household water savings (Albani et al., 2011).

During the last decade Jordan has undertaken major investments into the segment of waste water collection and treatment in all parts of the Kingdom. As reported by the MWI (unpublished) the average cost for the implementation of one meter of sewer canal has been 394 JD in 2009. The necessary investment assumed for the Wadi Shueib scenarios was inferred from these current sewer connection and rehabilitation figures and scaled to the level necessary to reach the anticipated sewer connection rates in As-Salt and Fuheis.

Decentralized Waste Water Treatment shows a large potential for implementation in Jordan (van Afferden et al., 2010). In a pilot study a working group in the SMART project developed investment scenarios for the villages of Ira and Yarka which were basically adopted for the planning scenarios.

Although Leakages from domestic cesspits and septic tanks are perceived as major threat to the groundwater in the study area (cf. Chap. 3.2.9) and the national water strategy explicitly states the objective to rehabilitate cesspits and enforce strict standards, there are no figures available on projects in this direction. For the FI scenario it was assumed that that the number of leaking tanks is reduced according to the national objectives and the rehabilitation costs per item are based on Sorge et al (2008) and scaled to the necessary figures.

In addition to the above mentioned investment estimated that were entirely based on the planning objectives of the national water strategy, the expected development in the study area demands two further future investments:

In order to achieve the stated supply objectives an increase of water imports appears necessary which would require the extension of the respective pipeline capacities. The expected costs for the respective pipelines and water mains were taken from (GTZ et al., 2009). Furthermore, increasing volumes of collected waste water will require an extension of the treatment capacities of the central plants in As-Salt and Fuheis. However, without a dedicated planning and design study it is not possible to provide satisfactory investment estimations. Thus, only a rough estimate for this study was inferred from the construction (or extension) cost data of 20 waste water treatment plants that have been built in Jordan during the last 20 years. Average construction costs of comparable plants were 2.5 JD per m³ treatment capacity (within a range of 0.5 to 5.7 JD per m³). This was taken as basic cost assumption for the extension of the treatment capacity in the study area.

Table 3.18 shows a summary of the investment assumptions for the Wadi Shueib planning scenarios as discussed above.

Table 3.18: Investment assumptions for the Wadi Shueib planning scenarios. The capital costs were assumed to be spread over the respective years and annualized according to their life expectancy (LE). A detailed discussion is given in the text.

Scenario:	Reference		BAU		FI		LE
Segment:	*JD*	*year*	*JD*	*year*	*JD*	*year*	*Years*
Water production	-	-	-	-	2,167,000	2011-13	25
Water import	500,000	2018-19	750,000	2014-15	250,000	2022-23	30
Supply network	-	-	1,961,775	2009-15	32,549,971	2009-24	(50)
Demand Management	-	-	-	-	1,458,333	2012-15	15
Sewers	-		26,457,000	2009-24	37,830,000	2009-24	40
Treatment	3,000,000	2015-16	9,000,000	2015-18	15,000,000	2015-18	30
DWWT	-	-	-	-	3,000,000	2015-24	30
Cesspits	-	-	-	-	1,575,000	2012-24	30

3.4.5. Scenario Simulation Results and Discussion

The following section presents the simulation results of the Wadi Shueib WEAP model for the four planning and the reference scenario. The results are presented and discussed on the basis of the performance indicators elaborated in Chapter 3.4.2.

In order to assess the influence of the variable annual water availability the results are discussed within the range of the 3 consecutive water years

- **2022/23** with relatively little rainfall (~84 % of the time series average)
- **2023/24** with relatively high rainfall (~120 % of the time series average)
- **2024/25** with average rainfall (~101 % of the time series average)

The development in comparison to the current situation is related to the water year **2008/09** with average rainfall (~112 % of the time series average).

3.4.5.1.　I.1a/b: Municipal Waste Water Treatment / Recharge Ratio

Following the model assumptions for water consumption, sewer connectivity and waste water leakages from canals and septic tanks (cf. Table 3.17) the ratio of collected and treated waste water in the Wadi Shueib area amounted to 75 % of the total discharged waste water volume in 2008/09. The scenario simulations for the waste water treatment ratio are shown in **Table 3.19** and **Fig. 3.35a**. A continuation of the sanitation project in As-Salt at the current rates (BAU scenarios) increases the ratio of treated waste water to 83 % until 2025. In the FI-scenarios the treatment ratio reaches 98 % until 2025, but requires a doubling of the yearly implemented house connections (from 100 to 200 new conncetions) in As-Salt, a comparable project in the municipalities of Fuheis and Mahis, as well as the realization of decentralized waste water treatment solutions and/or the rehabilitation and control of existing cesspits and septic tanks.

Table 3.19: Development of the municipal waste water treatment ratio (Indicator I.1a:TR_{MWW}) in the Wadi Shueib planning scenarios.

Scenario	Reference	LRP/HRP-BAU	LRP/HRP-FI
Year		%	
2008/09	75	-	-
2024/25	75	83	98

With the expected development of municipal consumption the absolute volume of untreated waste water that is discharged into the environment could increase significantly if no further efforts towards improved and controlled sanitation are taken. As depicted in **Table 3.20** and **Fig. 3.35**b/c the WEAP model estimates a current (2008/09) volume of approximately 1 MCM/a of untreated waste water discharge in the Wadi Shueib area which amounts to 8 % of the estimated groundwater recharge in the catchment. In a continuously dry period with high population growth (HRP scenarios), the discharged volumes are expected to constantly increase even under the current implementation strategy (HRP-BAU) to reach 1.6 MCM/a in 2025 and 10-15 % of absolute groundwater recharge, respective to wet (2023/24) or dry year (2022/23) conditions. Under the assumption of a future of reduced resources pressure (LRP) the BAU actions appear to at least stabilize the current figures. With the comprehensive FI-scenario actions the untreated waste water discharges are simulated to be effectively mitigated to levels below 0.25 MCM/a or 3 % of groundwater recharge, even in very dry years and with high population pressures (HRP-FI in 2022/23).

Table 3.20: Absolute volume of untreated waste water discharged in the study area (in MCM/a) and its ratio to simulated groundwater recharge (Indicator I.1b: $WW_{recharge}$; in %) for the year 2008/09 and the planning scenarios for the water years 2022-2025.

Scenario:	Reference		LRP-BAU		HRP-BAU		LRP-FI		HRP-FI	
Year	MCM/a	%	MCM/a	%	MCM/a	%	MCM/a	%	MCM/a	%
2008/09	1.01	8	-	-	-	-	-	-	-	-
2022/23	1.58	12	1.17	10	1.61	15	0.17	2	0.24	3
2023/24	1.63	9	1.17	7	1.63	10	0.12	1	0.16	1
2024/25	1.68	11	1.17	8	1.65	13	0.08	1	0.11	1

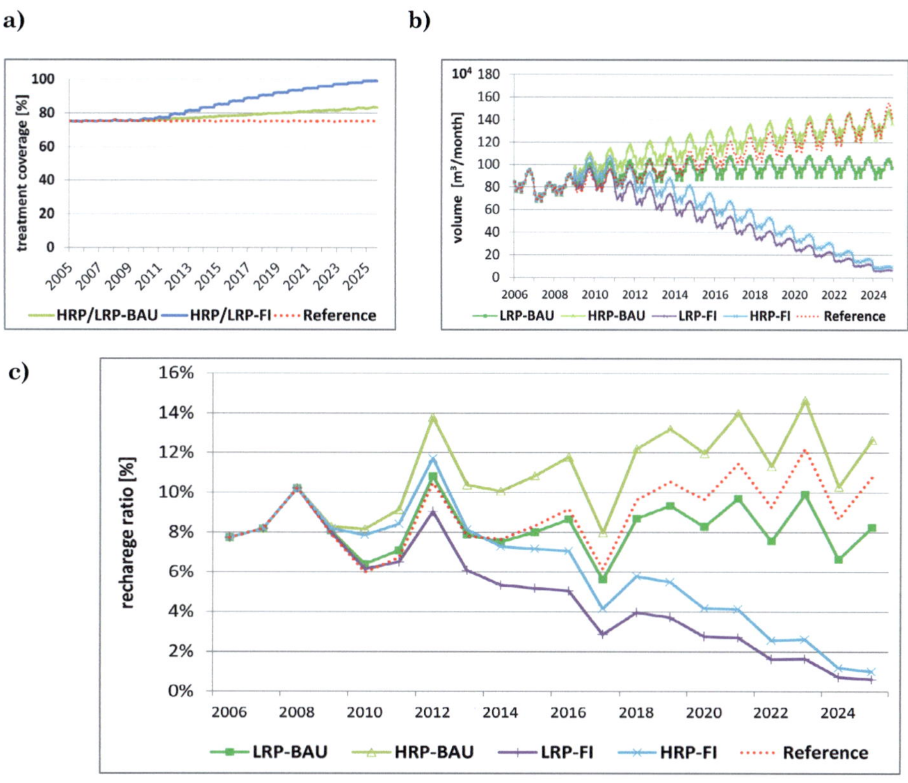

Fig. 3.35: (a) Municipal waste water treatment ratio (**Indicator I.1a:** TR_{MWW}); (b) Volume of untreated waste water in the catchment area; (c) Ratio of untreated waste water to absolute groundwater recharge (**Indicator I.1b:** $WW_{recharge}$).

3.4.5.2.   I.2. Available Renewable Water for Internal Use

Groundwater (+ imported freshwater)

For the water year 2008/09 the total estimated groundwater production in the Wadi Shueib area is 6.1 MCM or 49 m³ per capita, including spring and well production for municipal supply, as well as for industrial use and agriculture. The internal use after export and physical supply losses is estimated at 4.3 MCM or 33 m³ per capita. **Table 3.21** and **Fig. 3.36**a/b show the development of these figures in the scenario simulations.

Recovery of industrial production, increasing municipal demand and enhanced spring water quality due to the sanitation projects lead to an increase of produced groundwater to volumes between 8.7 (LRP-BAU) and 11.3 MCM (HRP-FI) in 2024/25 (**Fig. 3.36**a). The per capita availability of the internal groundwater resources (without imports) only increases noticeably in the FI scenarios in the range of 24-52 % depending on the scenario drivers and water year conditions.

The total per capita available internal resources (inclusive freshwater imports) increase in all scenarios from 50 m³ per capita in 2008/09 towards a considerable range from 52.3 to 78.1 m³ depending on the assumed demand development and the loss reduction in the scenarios (**Table 3.21**). As the model sets no restriction on import volumes, which thus basically fill the "demand gap" that cannot be covered by internal resources, the production increases are inversely related to the simulated water imports (**Fig. 3.36**b). The import ratio thus increase from 34 % of available drinking water resource in 2008/09 until 2023-2025 to 50-53 % (~32 m³ per capita) in the reference scenario and to 57-59 % in HRP-BAU (~45 m³ per capita). In the FI scenarios the imports can be reduced to 14-18 % (LRP) or stabilized at the current rates 34-38 % (HRP).

Table 3.21: Per capita volume of internal groundwater resources (and imported freshwater) available for use in the study area (Indicator I.2: $AWR_{internal}$). Current status for the year 2008/09 and simulated planning scenarios for the water years 2022-2025.

Scenario:	Reference		LRP-BAU		HRP-BAU		LRP-FI		HRP-FI	
	m³/c/a									
Year	*incl. import*	*excl. import*	*incl. import*	*excl. import*	*incl. import*	*excl. import*	*incl. import*	*excl. import*	*incl. import*	*excl. import*
2008/09	49.6	32.7	-	-	-	-	-	-	-	-
2022/23	57.7	27.9	59.3	32.5	72.2	30.4	52.3	45.3	65.2	43.3
2023/24	58.1	27.2	58.9	31.3	73.3	29.8	52.9	43.4	65.9	40.7
2024/25	63	31.2	65.3	36.1	78.1	33.3	57.7	49.6	70.5	45.8

Stored surface runoff and use

In comparison to groundwater, surface water exploitation plays a minor role in the study area, but is significant for downstream agricultural uses. The only direct surface water use modelled in the study area is the agricultural activity along the wadi channel with an estimated abstraction volume of approximately 0.6 MCM or 4.8 m³ per capita in 2008/09. **Table 3.22** and **Fig. 3.36**c/d show the development of these figures in the scenario simulations.

The simulated annual dam storage volume (**Fig. 3.36**c) is primarily dependant on the projected precipitation volumes (LRP or HRP). Furthermore, the increased utilization (production) of spring water and the consequent baseflow drop generates reduced surface runoff in the FI scenarios. Increasing irrigation demands in the HRP scenario result in an increase of internal stream water use to 1.2 – 1.3 MCM (6.6 – 7 m³ per capita) for 2022-2025 in the HRP-BAU scenario. Increased use of reclaimed water in the FI scenarios on the other hand result in a drop of internal surface water use for irrigation.

Table 3.22: Per capita volume of internal surface water resources utilized in the study area (Indicator I.2: AWR_internal). Current status for the year 2008/09 and simulated planning scenarios for the water years 2022-2025.

Scenario:	Reference	LRP-BAU	HRP-BAU	LRP-FI	HRP-FI
Year	$m^3/c/a$				
2008/09	4.8	-	-	-	-
2022/23	4.9	4.9	6.8	3.2	4.9
2023/24	4.7	4.7	6.6	3.2	5.3
2024/25	5	4.6	7	2.9	5.8

Reclaimed Water

Similar to surface runoff, the current use of reclaimed water in the study area is marginal, as only small plots are directly irrigated with effluent from the waste water treatment plants and some additional utilization can be ascribed to the use of mixed effluent along the wadi channel. The model estimates that in in 2008/09 a volume of $0.7\ m^3$ per capita was used internally while $17\ m^3$ per capita are exported as mixed effluent to the downstream agriculture. **Table 3.23** and **Fig. 3.36**e/f show the development of these figures in the scenario simulations.

Table 3.23: Per capita volume of internal reclaimed water utilized in the study area (Indicator I.2: AWR_internal). Current status for the year 2008/09 and simulated planning scenarios for the water years 2022-2025.

Scenario:	Reference	LRP-BAU	HRP-BAU	LRP-FI	HRP-FI
Year	$m^3/c/a$				
2008/09	0.7	-	-	-	-
2022/23	0.6	0.8	0.8	2.4	2.6
2023/24	0.5	0.7	0.8	2.6	2.8
2024/25	0.6	0.8	0.8	2.8	3

In conjunction with municipal consumption and increased treatment coverage the total volume of reclaimed water increases in all scenarios from 2.2 MCM/a in 2008/09 to 4.5 MCM/a (LRP-BAU) or 7.8 MCM (HRP-FI) in 2025 (**Fig. 3.36**e). In the FI scenarios the intensified utilization of reclaimed water for irrigation leads to an increase of internal use towards 0.5 MCM/a (~3 m³ per capita) in the period from 2022-2025. However, as agricultural demand is assumed not to develop strongly in the study area the increases in availability constantly outpaces utilization. Thus, the volume of treated effluent in the wadi stream is considered to increase significantly (**Fig. 3.36**f).

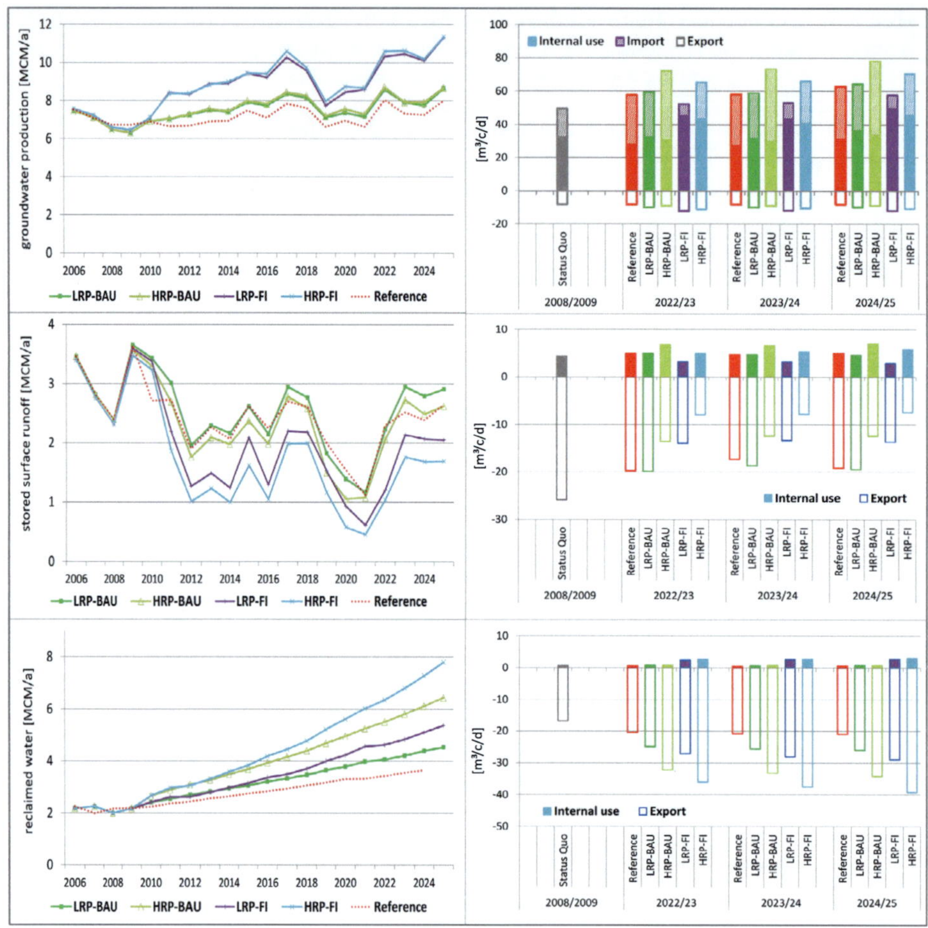

Fig. 3.36: Available (developed) water resources and use in the Wadi Shueib study area for (a, b) groundwater, (c,d) stored surface runoff (exclusive waste water treatment plant outflows), (e, f) reclaimed water.

3.4.5.3.   I.3: *Water Supply Shortage Index*

<u>**Municipal Supply Shortage Index**</u>

Since the Wadi Shueib WEAP model sets no restriction on import volumes, the water supply shortage is addressed here with regard to internal resources only. Water shortage are assessed in absolute terms [MCM/a] and as water supply shortage index (WSI) which ranges between 0-100 (cf. Chap. 3.4.2.3).

Unmet municipal demand (and consequently drinking water import) is estimated at 2.1 MCM and with a WSI of 21 for the water year 2008/09. **Fig. 3.37**a shows the development of these figures in the scenario simulations, **Table 3.24** and **Fig. 3.37**b shows the respective water imports necessary to fill the demand gap.

The simulation results for the unmet demand at the planning horizon range from a mitigated demand gap of 1.1 MCM/a (LRP-FI/wet water year) to a considerable increase of 8.3 MCM/a (HRP-BAU/average water year). In terms of the total annual volume of unmet demand the LRP-BAU (~4.6 MCM/a) and the HRP-FI (~4.3 MCM/a) scenario show a comparable performance to the reference scenario (~5.1 MCM/a) where no action at all is assumed. On the other hand the further reduction of the WSI of about 25-52 % indicates a reduced frequency of monthly supply shortages for both cases. The HRP-BAU scenario even performs far worse in comparison to the reference figures. According to these simulation results it appears that the FI strategy is the only way to secure a maintainable water supply for the growing demands in the study area (even under the LRP conditions). Correspondingly, the necessary water imports for municipal supply show a considerable range from 1.7 MCM/a (LRP-FI) to 8.7 MCM/a (HRP-BAU) for the year 2025.

Table 3.24: Unmet municipal demand [MCM/a] and annual water supply shortage index [-] (Indicator I.3: WSS_{DS}) in the Wadi Shueib area. Current status (2008/09) and simulated results for the planning scenarios for the water years 2022-2025.

Scenario:	Reference		LRP-BAU		HRP-BAU		LRP-FI		HRP-FI	
Year	MCM/a	[-]	MCM/a	[-]	MCM/a	[-]	MCM/a	[-]	MCM/a	[-]
2008/09	2.1	23	-	-	-	-	-	-	-	-
2022/23	4.9	38	4.6	29	7.8	44	1.6	3	4.5	16
2023/24	5.1	40	4.4	30	7.4	45	1.1	6	3.9	20
2024/25	5.4	42	4.8	32	8.3	48	1.4	6	4.6	21
2009-2025		25		22		29		11		17

Agricultural Supply Shortage Index:

Although agricultural demand for irrigation water has minor dimensions in the study area, there is nonetheless a competition for the available surface runoff with the downstream users, regulated by the traditional water rights. Especially in the dry summer months when irrigation demands are high and stream flows are low a demand-supply gap is expected.

It is estimated that the current unmet demand in the study area is about 1.3 MCM/a, with a WSI of 24 (2008/09). **Table 3.25** and **Fig. 3.38** show the development of these figures in the scenario simulations. Due to the increasing consumption and the resultant high volumes of reclaimed water and treated effluent the existing agricultural supply shortages are expected to become mitigated in the scenario simulation years.

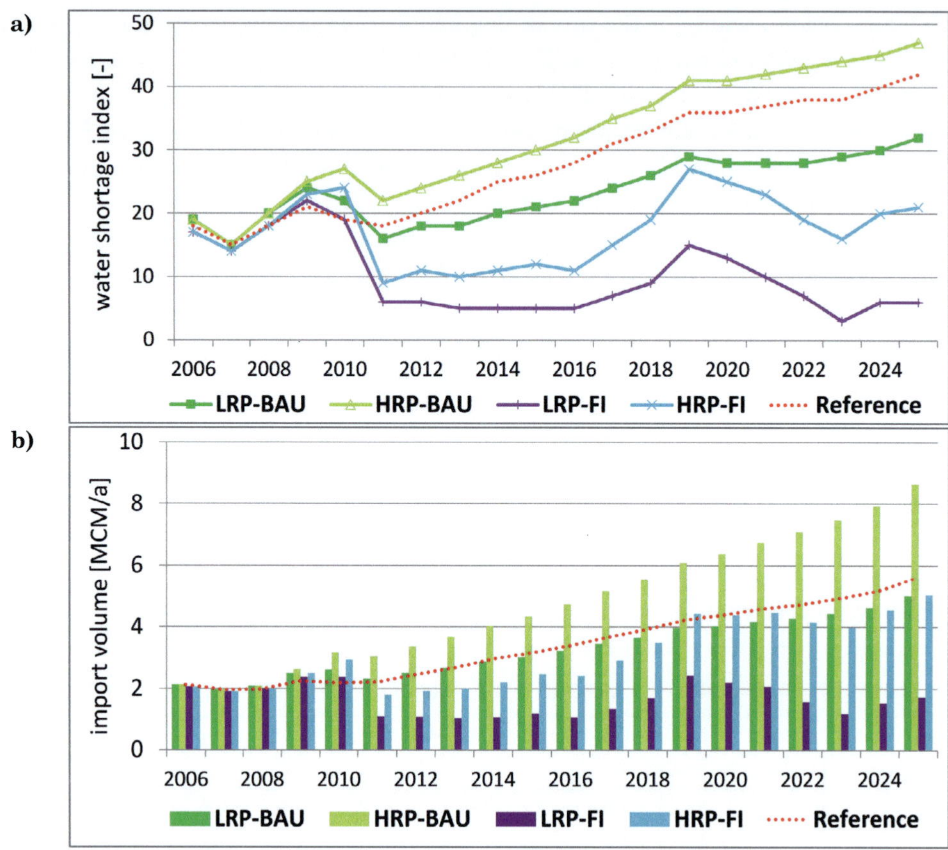

Fig. 3.37: Development of the municipal water shortage index (Indicator I.3: WSI$_{\text{MUNICIPAL}}$) and the respective water imports needs.

Table 3.25: Unmet agricultural demand [MCM/a] and annual water supply shortage index [-] (Indicator I.3: WSS$_{\text{DS}}$) in the Wadi Shueib area. Current status (2008/2009) and simulated results for the planning scenarios for the water years 2022-2025.

Scenario:	Reference		LRP-BAU		HRP-BAU		LRP-FI		HRP-FI	
Year	*MCM/a*	*[-]*	*MCM/a*	*[-]*	*MCM/a*	*[-]*	*MCM/a*	*[-]*	*MCM/a*	*[-]*
2008/09	1.3	*24*	-	-	-	-	-	-	-	-
2022/23	1.2	*12*	0.8	*7*	1.4	*11*	1	*14*	1.3	*11*
2023/24	1.3	*25*	0.9	*18*	1.5	*19*	0.9	*17*	1.3	*14*
2024/25	0.5	*6*	0.2	*3*	0.7	*5*	0.1	*3*	0.5	*3*
2008-2025		21		20		20		23		23

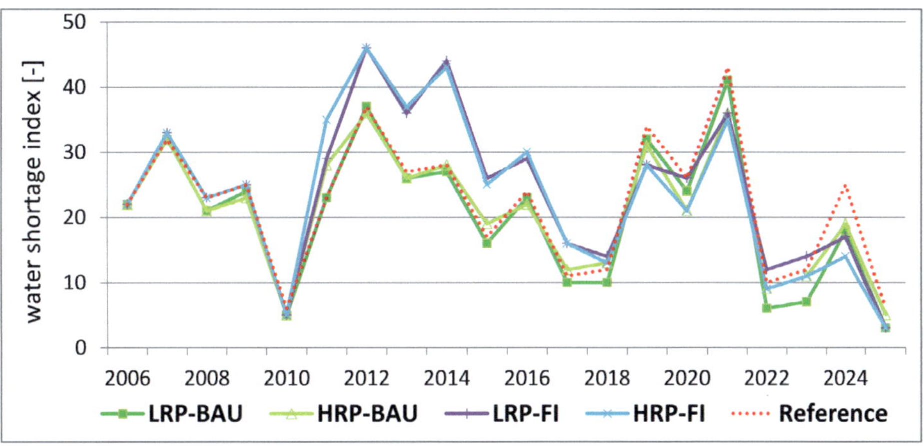

Fig. 3.38: Development of the agricultural water supply shortage index (Indicator I.3: WSI$_{AGRICULTURE}$) in the Wadi Shueib area for simulated planning scenarios.

Industrial Supply Shortage Index:

The industrial demand in the Wadi Shueib WEAP model (represented by Lafarge Jordan Cement) is self-providing from private groundwater abstractions. Thus, there is no supply shortage simulated for this sector.

3.4.5.4.   I.4: Water Supply Requirement

As an indicator for the efficiency of demand management and loss reduction measures in relation to growing demands, the water supply requirement in the Wadi Shueib has its primary focus on the municipal sector.

The total municipal supply requirement in 2008/09 for the municipal water demand is estimated at 10.0 MCM, corresponding to 56 m³ per capita. **Table 3.26** and **Fig. 3.39** show the development of these figures in the scenario simulations. Total municipal supply requirement is continuously increasing in all scenarios. In the HRP-BAU scenario the drinking water supply would need an upsurge by 113 % in order to meet the ambitious supply objectives of the water strategy, whereas under the LRP-FI assumptions increased supply needs amount to 24 %. On the other hand, in all but the reference scenario the supply efficiency (ratio of demand and required supply) improves from the current level of 71 % to 81 % (BAU) and 88 % (FI). This results in a more stable development of the per capita supply requirements in comparison to the total demand. The LRP-FI scenario is hereby the only setting succeeding in a reduction of per capita supply requirements to a relatively stable 51.7 m³ per person (~142 litres per day).

151

Since no distinct demand management measures are considered for the agricultural activities, the supply requirement in this sector develops according to the driving forces (cf. Chap. 3.4.4.4). Much the same applies for the industrial demand with the exception of the assumed implementation of a rainwater harvesting unit which reduces annual demand in the FI scenarios about 0.05 MCM in comparison to the related BAU scenarios.

Table 3.26: Water supply required for meeting the anticipated demand in the respective sector (Indicator I.4: WSR). The figures are given as total demand [MCM/a] and as demand per activity unit [m³ per capita/ton of industrial production/hectare of irrigated area].Current status (2008/2009) and simulated results for the planning scenarios in 2025.

Scenario:	Reference		LRP-BAU		HRP-BAU		LRP-FI		HRP-FI	
Municipal:*	*MCM/a*	*m³/c/a*	*MCM/a*	*m³/c/a*	*MCM/a*	*m³/c/a*	*MCM/a*	*m³/c/a*	*MCM/a*	*m³/c/a*
2008/09	10.0	56.4	-	-	-	-	-	-	-	-
2024/25	16.9	70.5	15.2	63	21.3	80.8	12.4	51.7	17.6	67
Industrial:	*MCM/a*	*m³/t*	*MCM/a*	*m³/t*	*MCM/a*	*m³/t*	*MCM/a*	*m³/t*	*MCM/a*	*m³/t*
2008/09	0.8	340	-	-	-	-	-	-	-	-
2024/25	1.0	340	0.9	340	1.0	340	0.8	323	0.9	323
Agricultural:	*MCM/a*	*m³/ha*	*MCM/a*	*m³/ha*	*MCM/a*	*m³/ha*	*MCM/a*	*m³/ha*	*MCM/a*	*m³/ha*
2008/09	2.7	675	-	-	-	-	-	-	-	-
2024/25	2.7	675	2.4	675	3.3	825	2.4	675	3.3	825

** including exports to Zai and Hashimieh areas*

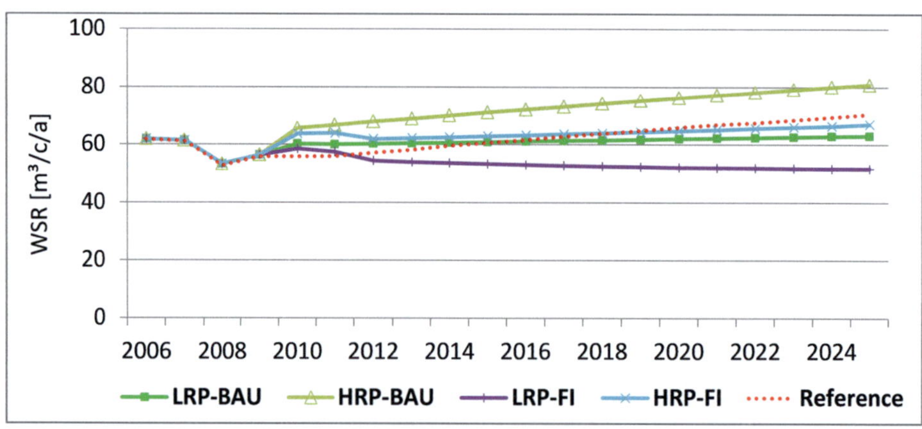

Fig. 3.39: Municipal water supply requirement (Indicator I.4: WSR).

3.4.5.5. I.5: Full Water Service Cost

As a relative measure for the economic cost efficiency of the different implementation levels the total and unit cost of full water service, including production, distribution, return flow collection and treatment, was calculated for the municipal water supply with regard to the necessary investments (cf. Chap. 3.4.4.5). Basic assumptions included:

- Projected demand is always fulfilled, whereby local resources are prioritized before water imports that supply remaining gaps.
- Operation and maintenance costs remain at their current level with exception of the special investments in the supply network maintenance segment (cf. Chap. 3.4.4.5).
- Current water costs are assumed without annualized capital costs from earlier investments.
- The annual interest rate is assumed at 5 %.
- Energy costs were assumed to remain at the subsidized level the WAJ is currently paying (43 fils per kWh).

For the water year of 2008/09 the full cost of municipal water services in the Wadi Shueib area was estimated at 3.33 million JD, or 0.552 JD per m³, whereof the production and supply part accounted for 2.47 million JD or 0.409 JD/m³. **Table 3.27** and **Fig. 3.40** show the development of the service costs after the investment of 3.5 million JD (Reference), 38.2 million JD (BAU) or 93.8 million JD (FI) until 2025 (cf. Chap. 3.4.4.5).

Table 3.27: Full water service costs for the municipal sector as unit cost in JD/m³ (Indicator I.5: FSC$_{UC}$) and total cost in million JD [MJD] per year. Current status (2008/09) and simulated results for the planning scenarios in 2024/25.

Scenario:	Reference		LRP-BAU		HRP-BAU		LRP-FI		HRP-FI	
Supply	[JD/m³]	[MJD]	[JD/m³]	[MJD]	[JD/m³]	[MJD]	[JD/m³]	[MJD]	[JD/m³]	[MJD]
2008/09	0.409	2.47	-	-	-	-	-	-	-	-
2024/25	0.449	4.38	0.395	4.04	0.428	5.88	0.489	4.68	0.483	6.16
Sanitation										
2008/09	0.143	0.86	-	-	-	-	-	-	-	-
2024/25	0.146	1.43	0.306	3.13	0.256	3.52	0.469	4.48	0.383	4.89
Full Service										
2008/09	0.552	3.33	-	-	-	-	-	-	-	-
2024/25	0.595	5.81	0.702	7.17	0.684	9.39	0.958	9.16	0.867	11.05

Total service costs rise in all scenarios to stabilize between 5.8 (Reference) to 11.1 (HRP-FI) million JD, whereby especially the sanitation segment contributes strongly to increasing the increasing costs, whereas the supply side costs are mitigated by some negative feedback through the reduction of water losses and expensive imports. The same effects also contribute to the relative stabilization of the unit costs which reach 0.6 JD/m³ in the Reference, 0.68-0.70 JD/m³ in the BAU and 0.87-0.96 JD/m³ in the FI scenarios. Since the extension of waste water treatment capacities were assumed equal for the LRP and HRP scenario branches, the HRP branch attains a better utilization rate, thus resulting in lower unit costs of sanitation. Therefore, the full service price is estimated as highest in the LRP-FI scenario.

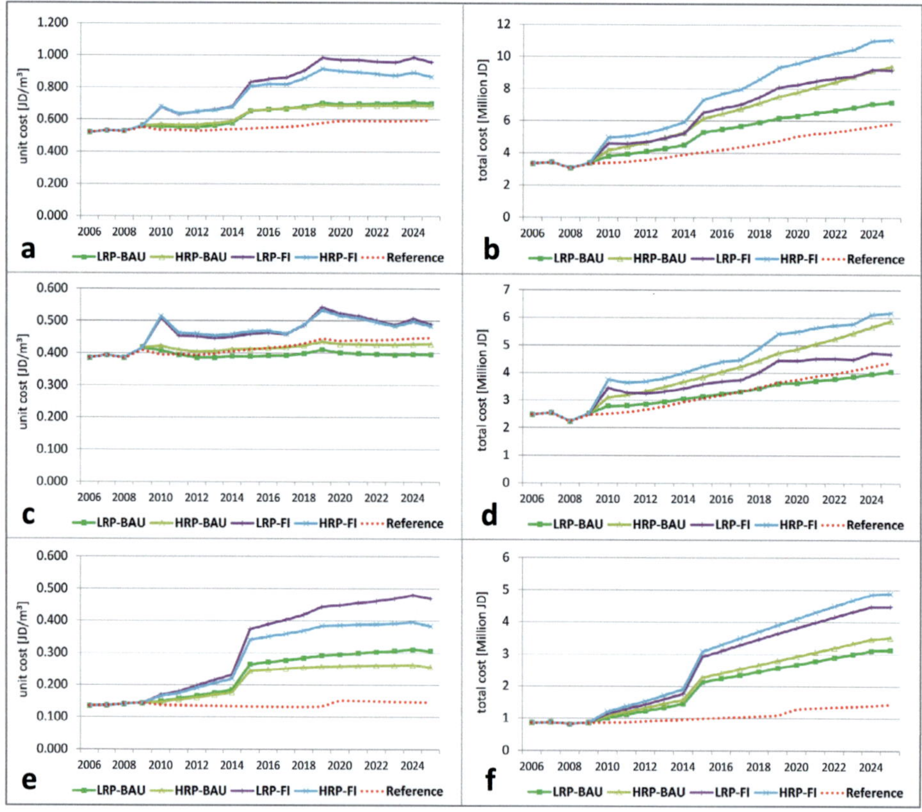

Fig. 3.40: Full service cost of municipal water as **(a)** unit cost in JD/m³ and **(b)** total costs in million JD per year. The full service costs include production and supply **(c+d)** as well as sanitation services **(e+f)**.

3.4.5.6.   *I.6: Environmental Water Stress*

Environmental water flow requirement in the Wadi Shueib was assessed on the basis of the natural flow conditions according to the historical records and the water balance (cf. Chap. 3.3.6 and 3.4.2.6). Natural mean annual runoff, including baseflow and flood flow, is estimated to amount 13.5 MCM. With a Q90-monthly-flow of 0.27 MCM and an inferred high flow requirement of 2.69 MCM the environmental water requirement (EWR) is estimated at 2.97 MCM/a.

According to the balance figures for the period 2001 until 2009 the environmental water stress indicator under consideration of return flows (WSI_{RF}) of the Wadi Shueib stream ranges between 0.7 and 1.0. This indicates a continuous medium to high water stress with a margin of 0-30 % of stream flow volume left to drop below the EWR and into overexploitation and environmental water scarcity.

Fig. 3.41 depicts the temporal development of the WSI_{RF} in the scenario simulations. The WSI_{RF} remains within the range of the water stress category ($0.6<WSI_{RF}>1.0$) neither showing a significant trend towards overexploitation nor to water abundance. This basically means that the increasing exploitation of base flows is effectively balanced by the rising discharges of the waste water treatment plants (cf. **Fig. 3.36**). Furthermore, it appears that even during dry periods (e.g. 2012-2014 and 2019-2021) the more climate-unrelated treated effluent discharges prevent a dramatic drop into water scarcity conditions. On the other hand, very wet water year conditions (e.g. 2017-2018) temporarily mitigate the catchments water stress. Overall, due to stronger water utilization rates, the FI scenarios appear to perform slightly worse in terms of the WSI_{RF}.

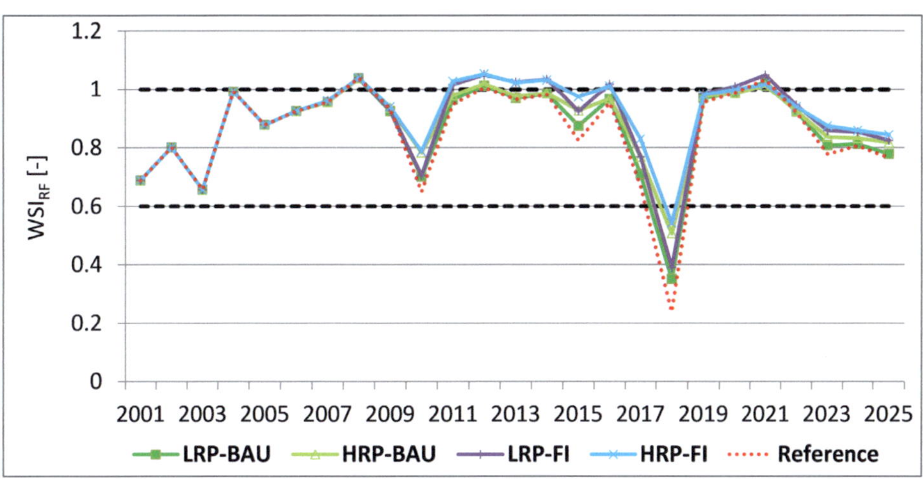

Fig. 3.41: Environmental water stress with consideration of return flows (Indicator I.6: WSI_{RF}) for the Wadi Shueib streamflow for the historical records 2001-2009 and the scenario simulations 2009-2025. The horizontal broken lines delineate the categories of water stress (above lower line) and water scarcity (above upper line).

3.4.6. Summary and Discussion of the Scenario Simulations

In order to facilitate a direct comparison of the performance of the scenarios it is necessary to standardize the simulated indicator values to a common scale. For that purpose different standardization techniques are proposed in the Multi-Criteria Decision-Analysis literature that either use linear or non-linear transformation or value/utility functions to transform the decision criteria to a common scale (see e.g. Malczewski, 1999). The selection of a suitable transformation procedure depends on the general decision process, the distribution of the criterion data and other factors and can result in different rankings of the decision alternatives (e.g. Young et al., 2010).

For the purpose of comparative summary of the scenario simulation results a simple *maximum score procedure* transformation was found useful. In order to achieve a common scale it has to be distinguished between benefit criteria (where maximizing the criteria value is desirable) and conversely cost criteria (where lower criterion values are desirable). The selected performance indicators in this study are all cost indicators with exception of *I.1a: municipal waste water treatment ratio* and *I.2: Available renewable water for internal use*, where obviously maximization is desired. In order to have scores with a common anchor at 0.0 (the scenario with the worst performance has always a score of zero for an indicator) the indicator score for the cost-type indicators is given by

$$x'_{ij} = 1 - \frac{x_{ij}}{x_i^{max}}$$

where x'_{ij} is the standardized score for the i-th indicator and the j-th scenario, x_{ij} is the original value and x_i^{max} is the maximum value for the respective indicator.

In the case of benefit-type indicators criterion the indicator score is given by

$$x'_{ij} = \frac{x_{ij}}{x_i^{max}} - \frac{x_i^{min}}{x_i^{max}}$$

where x'_{ij}, x_{ij}, x_i^{max} are as defined above and x_i^{min} is the minimum value for the respective indicator.

Table 3.28 shows the resultant standardized indicator scores in the evaluation matrix. The score range of the Reference, BAU and FI scenarios delineates the maximum and minimum of each indicator performance within the respective LRP and HRP scenario sets and within the period from 2023-2025 (dry, wet and average water years).

Table 3.28: Evaluation matrix of standardized indicator scores.

Indi-cator	Sub-indicator	Reference		BAU		FI	
		score range		*score range*		*score range*	
I.1	I.1a: WW treatment ratio	0.00	0.00	0.08	0.08	0.23	0.23
	I.1b: WW recharge ratio	0.20	0.40	0.00	0.53	0.80	0.93
I.2	I.2a: available groundwater	0.00	0.08	0.05	0.18	0.27	0.45
	I.2b: available surface water	0.26	0.30	0.24	0.59	0.00	0.41
	I.2c: available reclaimed water	0.00	0.03	0.07	0.10	0.63	0.83
I.3	I.3a: municipal shortage	0.13	0.21	0.00	0.40	0.56	0. 94
	I.3b: agriculture shortage	0.00	0.76	0. 24	0.88	0. 32	0.88
I.4	I.4a: municipal supply requirement	0.21	0.21	0.00	0.29	0. 17	0.42
I.5	I.5: unit cost	0.38	0.38	0.27	0.29	0.00	0.09
I.6	I.6: environmental water stress	0.08	0.08	0.03	0.06	0.00	0.02
	Total:	**1.26**	**2.45**	**0.72**	**3.40**	**2.49**	**4.26**

As expected the FI-alternative shows the overall best performance for most of the selected indicators and a total score in the range from 2.49 to 4.26 points. Under favourable conditions (LRP) the BAU alternative beats the Reference and the HRP-FI scenario with a total score of 3.40, but in the case of HRP conditions it shows the worst performance for most of the indicators and for the total score of 0.72 points. It has to be kept in mind that for the Reference scenario the external driving forces were assumed to develop on an intermediate path between the LRP and the HRP assumptions. Thus, a comparison and interpretation has to pay attention to this bias. Furthermore, due to the standardization procedure the indicator scores range between 0.0 and 1.0, this, however, does not mean that a value of 1.0 represents an absolute optimum, but rather a scenario with an indicator performance that is significantly higher in comparison to the scenario with the minimum performance.

In **Fig. 3.42** the performance scores for the most significant indicators for the planning scenarios is plotted for comparison. Especially in the issues of untreated waste water recharge (I.1b; cf. Chap. 3.4.5.1) and municipal water supply shortages (I.3a; cf. Chap. 3.4.5.1) the FI scenarios outperform the BAU alternatives significantly for both the LRP and HRP assumptions. It is also noteworthy that with regard to the untreated waste water recharge ratio (I.1b) the BAU scenario shows a comparatively wide range between LRP and HRP conditions, whereas the FI performs more stable. It has to be stated that, beyond the regularly recorded coliforms in the Wadi Shueib springs (cf. Chap. 3.2.9), there is yet no clear understanding of the processes and the amounts to which discharged waste water recharges the groundwater resources in the study area. Nonetheless, the simulated high volumes of untreated waste water in relation to the groundwater recharge estimates can certainly be understood as important and warning groundwater vulnerability factor, especially with regard to the karstified environment of the Wadi Shueib area.

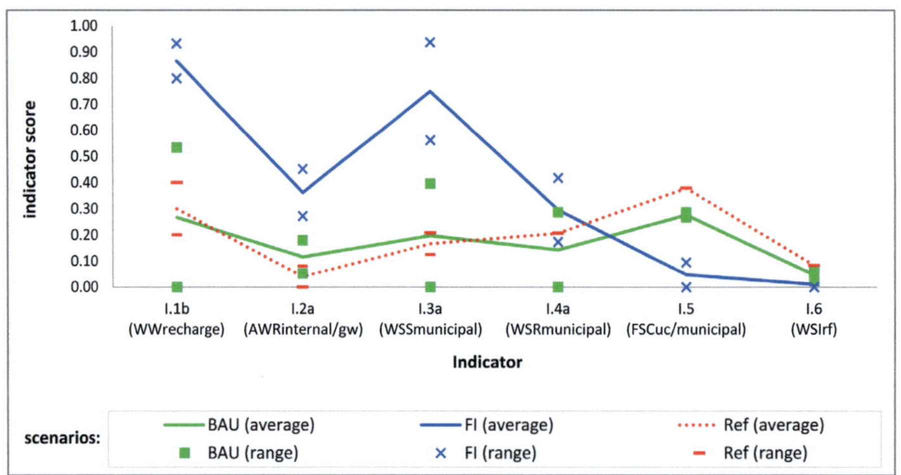

Fig. 3.42: Summary of the standardized scenario simulation results. Higher score means a better performance while range expresses sensitivity to the external scenario driving forces (LRP/HRP). For the indicator acronyms please refer to Table 4.2 and Chap.3.4.2.

In terms of readily available and utilizable freshwater (groundwater) within the catchment (I.2a; cf. Chap. 3.4.5.2), the investment in water production as well as the anticipated strong improvements of groundwater protection also leverages the FI above the BAU scenarios, whereas for the latter alternatives improvements of groundwater quality were also expected but of lesser extent. The scores for this issue display a smaller magnitude of improvement for the FI scenarios. But although given the fact that the possibility of developing new or maximizing available water resources in the study area is limited, this of course does not mean that the improvements are less substantial for the water sector in the study area. Surface water uses change slightly in the scenarios, but in relation to the amounts stored and used downstream the scenario implementations have no strong effect on surface water availability internal or external.

The total municipal water supply requirement (I.4a; cf. Chap. 3.4.5.4) is expected to rise in all scenarios, but the magnitude is strongly dependent on the driving forces of population growth and demand development. According to the simulations it appears likely that increased water imports become necessary if the supply objectives of the national water strategy are to be met. In this light, considerable investments in demand management, water infrastructure and the efficient reduction of water losses appear urgently necessary when these impacts are to be mitigated in the Jordanian water future. Similar recommendations are found in current studies on the status and future of the Jordanian water sector (McIlwaine, 2009; Rosenberg & Lund, 2009; Sommaripa, 2011).

With regard to the simulations of environmental water stress (I.6; cf. Chap. 3.4.5.6) all scenarios show a relatively similar performance ranging in the category of water stressed with fluctuations mainly dependent on the water year conditions. With a slight offset, the FI scenario shows the worst performance under both the LRP and HRP conditions, due to the high abstraction of the wadi base flow (spring discharge). It has to be kept in mind that the environmental water stress as discussed here merely addresses water quantity. Quality concerns and possible impacts on ecosystem functions, due to the continuously increasing share of treated effluent in the wadi runoff would need a separate assessment that could not be undertaken within this study.

3.5. Summary

The preceding chapters documented an exemplary modelling and scenario planning exercise for sub-basin-scale integrated water resources management with an application to the Jordanian Wadi Shueib catchment. Modelling was purposefully grounded on the data and information basis as it would currently be available for fully informed planning personnel and decision makers in the Jordanian water sector institutions (cf. Chap. 3.3.3). The data was subjected to a thorough data quality control (cf. Chap. 3.3.4) and several pre-processing steps, in order to generate consistent water balance time series, which allowed the estimation of important hydrological and resources engineering model assumptions (cf. Chap. 3.3.5).

The Water Evaluation and Planning (WEAP) model was used to represent the water system of the study area and act as IWRM planning tool (cf. Chap. 3.3.7). The WEAP model was calibrated and validated against the Wadi Shueib Reservoir data series and provided an overall good prediction of the total annual received water in the Wadi Shueib Reservoir (cf. Chap. 3.3.7.3). The major uncertainty in the model assumptions was identified as the transmission loss parameter (cf. Chap. 3.3.7.4).

National water strategy objectives and action plans were used as normative guideline to craft a set of planning alternatives with direct relation to the local challenges (cf. Chap. 3.4.1). The systematic selection and development of IWRM performance indicators was organized in direct reference to the stated objectives and the constraints inherent in the WEAP framework (cf. Chap. 3.4.2). For the scenario approach the identified alternatives were combined in two action plans for the Wadi Shueib area: a Business as Usual (BAU) strategy assuming the currently active projects are continued and finished until 2025 and a Full Implementation (FI) development line assuming the full range of stated implementation approaches is realized until 2025 (cf. Chap. 3.4.3). External driving forces developments were modelled to agree with the official projections for climatic conditions, demographic growth and water demand development. Internal drivers were modelled as rate of change and cost figures, based on a review of past performances and the respective current planning status of the Jordanian water sector institutions (cf. Chap. 3.4.4).

The scenarios were simulated for the elaborated performance indicators and discussed within the range of the 3 consecutive water years, constituting dry (84 % of average rainfall), average (101 %) and wet (120 %) year conditions and compared to the current situation in the water year 2008/09 (cf. Chap.3.3.6).

3.6. Conclusions

The IWRM modelling and scenario planning approach presented in this study documents a consequent and complete process of

- developing a holistic water system understanding based on a heterogeneous pool of available data in the Wadi Shueib area,
- developing a detailed WEAP IWRM model at the implementation scale in the context of the Lower Jordan Valley area,
- developing local planning alternative, objectives and suitable performance indicators on the basis of the national water strategy
- and providing an exemplary scenario simulation to show how the detailed local planning reality compares to national goals.

3.6.1. Water Accounting for the Wadi Shueib Catchment

The Wadi Shueib is regarded as a rather well studied area (cf. Chap. 3.2.1). Nonetheless, the available data at the water sector institutions, as well as the reviewed literature, revealed several uncertainties, critical knowledge gaps and disagreements regarding the water resources and related topics (cf. Chap. 3.3.3). In large parts, this is due to the often inconsistent monitoring of the institutions and the fact that previous studies primarily focused on specific and isolated aspects of the catchment balance.

Thus, the thorough processing of the available data basis from the water sector institutions, as well as the literature information, is the **first comprehensive analysis of the holistic Wadi Shueib water balance and a unique contribution to the understanding of the Wadi Shueib water resources**

system. In this respect, also the compiled water balance schemes (cf. Chap. 3.3.6) have been found a useful tool to keep track of the status and advances of the water system knowledge that can easily be transferred to other study areas.

Furthermore, **several points could be revealed, where further detailed studies are necessary to reduce uncertainties in the process understanding of the complex IWRM domain**. This already led to a series of consecutive studies by an interdisciplinary expert group in the SMART project which will help to continuously substantiate the presented IWRM model and scenario assumptions (Alfaro et al., 2012; Grimmeisen et al., 2012; Müller et al., 2011; Zemann et al., 2012).

3.6.2. The Wadi Shueib WEAP Model as an IWRM Modelling Tool

WEAP comes with a clear focus on basin scale water balancing and allocation scenarios. Rather generic modelling possibilities enable implementation of elements not directly provided by the user interface (e.g. springs and water treatment plants) as well as some scale independence. WEAP has therefore been found a flexible and ample tool, generally appropriate for water resources system modelling in the applied case study. **Despite the complexity of the assessed IWRM system and the detail of the sub-catchment scale, the Wadi Shueib WEAP model could be successfully constructed to represent all elements of the preliminary prepared holistic water balance schemes (cf. Chap. 3.3.7).** However, it also has to be kept in mind that the groundwater component in the catchment was represented with a relatively simple bucket model of a single cretaceous aquifer complex (A1-A7). Although WEAP can link to a MODFLOW model component, employing a numeric groundwater model appeared not feasible, due to the difficult and largely unknown geological situation in the study area and the absence of a suitable observation wells network. Furthermore, a set of general assumptions had to be made in the effort to find best model estimates necessary for the IWRM complexity. Also, calibration had to be conducted manually by iterative adjustment of the monthly parameter values after model runs, because the stepwise hydro-balance algorithm in WEAP does not cope with interdependent parameters within one time-step such as simulated water volume in reservoir and reservoir seepage.

Eventually, the final model displays good performance for the assessment of the monthly integrated catchment water balance (cf. Chap. 3.3.7.3). The peak storages of the Wadi Shueib Reservoir appear well simulated for all years except 2005/06. However, the calibrated transmission loss parameter, although in the range of comparable literature estimates, might aggregate various other estimation errors. More reliable logging of dam operation habits and discharge records could help to significantly reduce remaining uncertainties here. For the intended model use of long term IWRM scenario planning, however, these uncertainties were accepted.

Overall, it was found that WEAP can contribute an important part to IWRM planning processes with a focus on water balance and allocation, especially when it is well embedded in a comprehensive and soundly constructed IWRM scenario planning framework as demonstrated in this study. An important strength of WEAP was found to be the integrated scenario management that allowed a straightforward and suitable modelling of to the scenario assumptions undertaken here. **The final WEAP model is well documented in this thesis, as well as in the parallel developed knowledge platform DROPEDIA (cf. Chap. 4) and is ready to be employed and extended by the Jordanian decision making institutions.** Hereby, the free licensing for developing countries, the intuitive usability and the flexible nature are essential benefits of the WEAP approach. Owing to these advantages, several water sector institutions in the Lower Jordan Valley have recently already begun to adapt WEAP in their strategic planning directorates.

3.6.3. IWRM Planning in the Wadi Shueib

The current Jordanian Water Strategy (RCW & MWI, 2009) is an ambitious policy that comprehensively addresses the critical issues in its stated objectives. Furthermore, during the last decade tremendous efforts have been invested in the water sector by the Jordanian government as well as the international community. So far, however, these efforts were often not streamlined and sometimes mismanaged (Humpal et al., 2012). This often leads to the situation that important and intricately obtained information gets lost after projects are finished. In better cases, results are summarized into a report, but often the documentation is incomplete and the knowledge leaves the region with the project or the worker. In any case, it appears very difficult to keep this large pool of knowledge alive in the IWRM process.

Within the presented work, it was shown, that the actual planning practice only considers a small subset the comprehensive national goals and implementation objectives (cf. Chap. 3.4.1). Hereby, **the detailed analysis of the normative objectives with regard to necessary implementation options and suitable performance measures (indicators) is another important contribution to the IWRM planning endeavours in the Wadi Shueib area.** In this regard the study offers a proof of concept and a methodological framework for the application in other sub-basin scale IWRM planning applications.

The combined modelling and scenario planning approach also shows **a potential path towards an operational realization of the ambitious IWRM hypothesis on the sub-basin scale** and, thus, a step further towards realization. The modular and flexible nature of the presented framework appears applicable for expansion and continuous development

3.6.4. Results of the Scenario Simulation

The scenario simulations point to the conclusion that only an increased effort (FI scenarios) can enable a progress towards the goals of the national water strategy and a successful IWRM process in the Wadi Shueib area. **The currently undertaken projects in the area, even when prolonged into the near future (BAU scenarios) appear not to be able to relieve pressure from the water situation in many issues.** Furthermore, the rate of success for the issue of meeting future water requirements is strongly dependent on the uncertain development of near to mid-future climatic conditions, population growth and water demands for all implementation strategies addressed.

The major downside of the FI alternative is of course related to the considerable investment costs involved and the resultant rise of the water service costs. Due to the feedback of reducing expensive water import volumes and decreasing water losses the unit cost of water services does not rise linear with the anticipated investments. Yet, it was still simulated to significantly increase from the current $0.595\ JD/m^3$ to $0.958\ JD/m^3$ in the LRP-FI scenario in 2025 (cf. Chap 3.4.5.5). However, in the light of the current water sector expenditures (including MWI, WAJ, JVA, Aqaba Water Company, and Miyahuna) that were estimated at approximately JD 500 million in 2010 (Sommaripa, 2011), the anticipated investment costs for the FI scenarios (~JD 94 million over 15 years in a region with 2-3 % of the country's population) do not appear overambitious. Even more so with regard to aspiring mega-projects like the Red-Sea-Dead-Sea canal with currently estimated costs of approximately JD 4 – 8 billion (World Bank, 2009a).

In any way it is expected that another water tariff reform will be necessary in order to progress towards the objective of cost recovery in the water sector, since already today the revenues of the WAJ and the public companies do not cover their annual expenditures (Segura/IP3 Partners LLC, 2009).

The bottom line of the simulation results in the Wadi Shueib area might also hint at the general Jordanian situation. During the last decade the water sector has mostly focused on the maximization of available and the development of new resources (e.g. use of reclaimed water, brackish water desalination) as well as the extension of sanitation services. And even though remarkable accomplishments have been achieved in these segments, the water situation has not been eased. On the other hand, the reduction of the tremendous water losses and the realization of efficient demand management and water awareness campaigns have played secondary roles until now.

The established scenarios are of course only a limited set that span a basic framework. For the further application in planning and strategy decisions, the development of further scenarios is recommended where for example only certain aspects of the FI options are taken into consideration, or other combinations of driving forces are assessed. **The model presented in this study is flexible enough to implement such changes in relatively short time.**

4. Knowledge Management for IWRM Plannig and Decision Support

4.1. Introduction

The previous chapter demonstrated the concrete implementation of a holistic IWRM modelling and scenario planning concept for a sub-basin catchment of the Lower Jordan Valley. The combination of physical system modelling with social, ecological and economic factors required the integration of data and information from several different domains. The impacts of the planning scenarios were assessed from various perspectives, represented by an interdisciplinary set of indicators. Hereto, the author was building on the pool of previous works for the study area, as well as on the cooperation with sector experts and stakeholders from within the framework of the SMART project (Wolf & Hötzl, 2011).

Transferred to a productive and institutionalized environment, an IWRM analysis of this manner involves a highly collaborative process between such distributed knowledge sources. In this respect, some discussion has already been progressing in recent years, about how to span the gap between scientific environmental modelling and respective political decision making (Acreman, 2005; Cash et al., 2002; Kolkman et al., 2005; Liu et al., 2008). Most contributions on this topic agree on the urgent need to improve communication and transparency between science and policy. Regarding the multi-disciplinary nature of IWRM, it is also the information flow between the participating experts that poses challenges.

The currently established formalized communication channels for the knowledge transfer between experts and decision makers basically consist of reporting, presentation and scientific publication. The relevance of these instruments is not questioned and especially in terms of quality-control the process of peer-reviewed publishing is undisputed. On the other hand, new web-based communication tools like blogs or wikis largely improve collaboration potentials. Scientists that have become aware of such instruments praise the possibilities of a more continuous flow of information and feedback and more timely discussion between research participants (Butler, 2005). Another important incentive is seen in the possibility to document and share the many bits and pieces of work in the daily life of a researcher that actually never make it to a publication or a project report, in the best case are informally communicated within an active group of interest, but as often never leaf the sketchbook of the scientist. This so-called "file-drawer-problem" (Rosenthal, 1979) has been long discussed and various authors see a large potential to make science more effective by sharing such details that are often obscured by the classical communication mechanism (De Roure et al., 2010; Waldrop, 2008). Especially in the domain of laboratory research in life sciences and chemistry a young generation of scientists has eagerly commenced on utilizing online communication and collaboration technology to freely share their scientific opinion, data and knowledge in so called "Open Notebook" research, or more generally in the "Open Science Movement" (Bradley et al., 2008; Ekins et al., 2011; Williams, 2008).

Also under water professionals, the development of a comprehensive and interdisciplinary water resources knowledge base is perceived as fundamental to integrated water resources assessments and consequent planning and decision making (GWP, 2000). It is also recognised that contemporary knowledge management techniques can contribute to improving the performance and effectiveness of both capacity development (Luijendijk & Lincklaen Arriëns, 2007) and knowledge sharing (Giupponi & Sgobbi, 2008; Roux et al., 2006) in the water sector, provided that there is a basic capacity in place to coordinate this approach. To handle this necessity, most IWRM projects carried out during the last years already had knowledge sharing initiatives on their agenda. The requirement for such a knowledge framework is tackled in current IWRM approaches in several manners: (1) Applied projects and case studies usually share multi-thematic databases and Information Systems to enable data access, transfer and comparability between their interdisciplinary modelling environments. (2) Joint projects implement internet platforms for building and distributing thematic bibliographies as information portals on IWRM related topics, or (3) focus on supporting Capacity Building Networks in the IWRM environment.

Although growing in other disciplines, the direct support of collaborative knowledge-building and sharing processes has been less prominent in water research. Several organizations have recently started initiatives of which probably the most visible example is the UNDP-initiated WaterWiki (WaterWiki, 2012). But especially in the planning practice respective approaches are still lacking satisfactory consideration in state of the art approaches on IWRM-Decision-Support-Systems.

But for all that, the potential for a continuous online collaboration process appears evident as (1) the process is meant to be highly collaborative per se, (2) the community of experts often is spatially and disciplinary distributed and (3) planning and decision-making has often to be undertaken immediately and constantly and cannot stall until an eventual knowledge transfer process at some point in the future.

The following chapter presents the conceptual design and the implementation of a Semantic MediaWiki-based knowledge management platform to approach IWRM towards these purposes. Development and preliminary application was centred on the experience and insights from the Wadi Shueib IWRM model study. Yet, equal attention was given to an unhampered applicability to comparable case studies as well as to a structural generality that allows for a broad application potential in the general IWRM planning practice. The presented instrument was also designed to be eventually embedded into the decision support framework of the SMART research project (**Fig. 4.1**). The other components include a central database, various expert models and detail studies as well as modules for Multi-Criteria Decision Analysis and Optimization (Wolf & Hötzl, 2011).

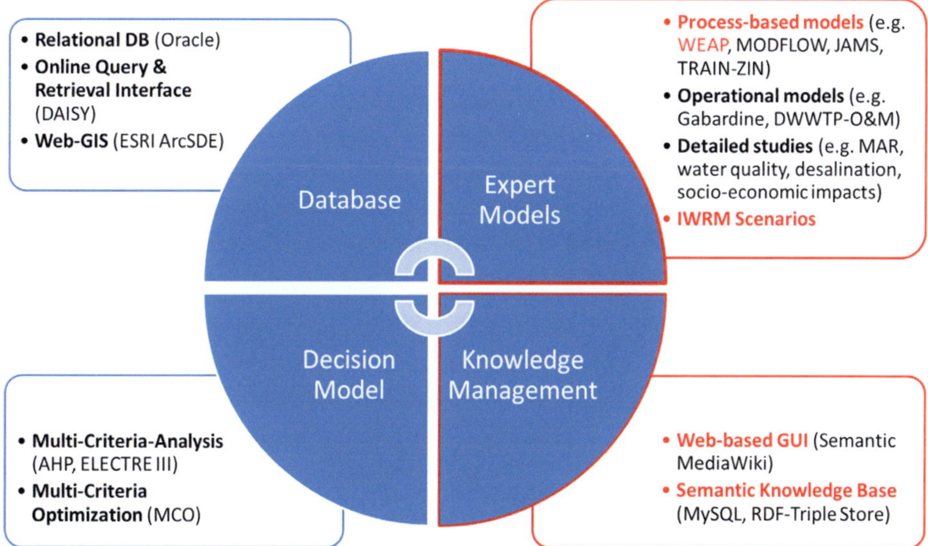

Fig. 4.1: Components of the SMART-Decision Support framework for Integrated Water Resources Management in the Lower Jordan Valley. Elements marked in red are investigated and discussed in this thesis.

4.2.　Perception of KM in the Jordan Valley Water Sector

Any knowledge management initiative stands and falls with the willingness to contribute of its addressed users. In order to assess the basic perception on the subject among water professionals associated to the study region, a set of questions was given to the attendees of a large SMART coordination meeting in 2011 (Wolf et al., submitted). A total of 53 participants (38 % from Jordan, Palestine and Israel, 62 % from Germany), including scientists from the project consortium and associated stakeholders, filled out the anonymous questionnaire. **Fig. 4.2**a-d displays the outcomes specifically addressing knowledge management issues.

On a general level, a strong majority agrees that more publicly available information would increase the public acceptance of water policy, both on transboundary and local level (**Fig. 4.2a**). Transparency appears to be perceived as important facilitator in a political decision making that follows the national guidelines. However, in terms of current stakeholder participation processes, opinions are far more ambiguous. Especially on the local level, the majority is rather undecided whether stakeholder participation is sufficiently supported.

The participating experts generally responded positively (72 %) on the proposal of using a wiki based platform for support IWRM in the context of their work and on the time they were willing to invest into such an initiative (**Fig. 4.2b**). Again, transparency and public information appear as strongest incentives for contribution, followed by information retrieval and dissemination on a professional level (**Fig. 4.2c**).

Taken together, these opinions point towards a strong feeling for the need of improved communication and information between decision makers, water sector experts and the wider public, but at the same time water professionals are cautious about facilitating direct participation in decision processes (a possible flaw in the questions might be related to the unclear definition of 'stakeholder'). Thus, following this opinion poll would recommend a semi-open information and participation system.

a)

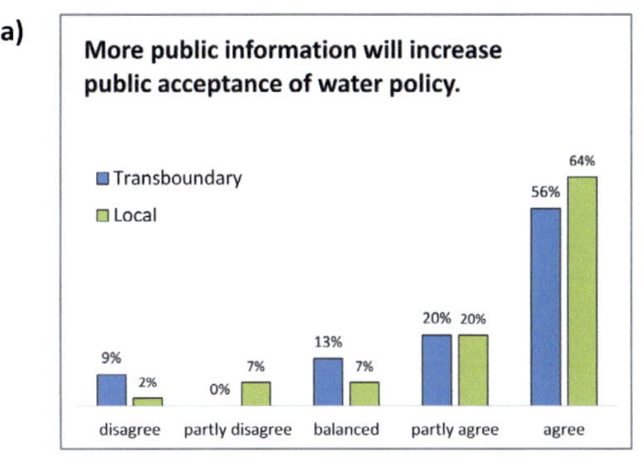

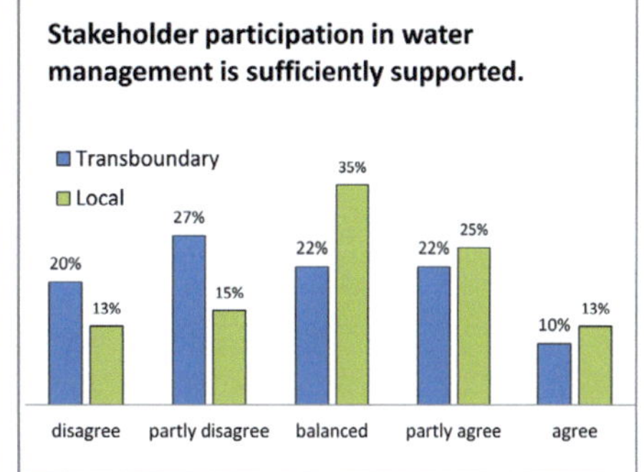

b)

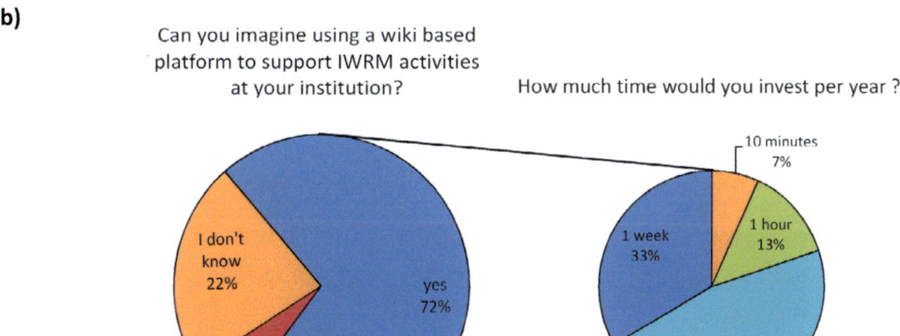

c)

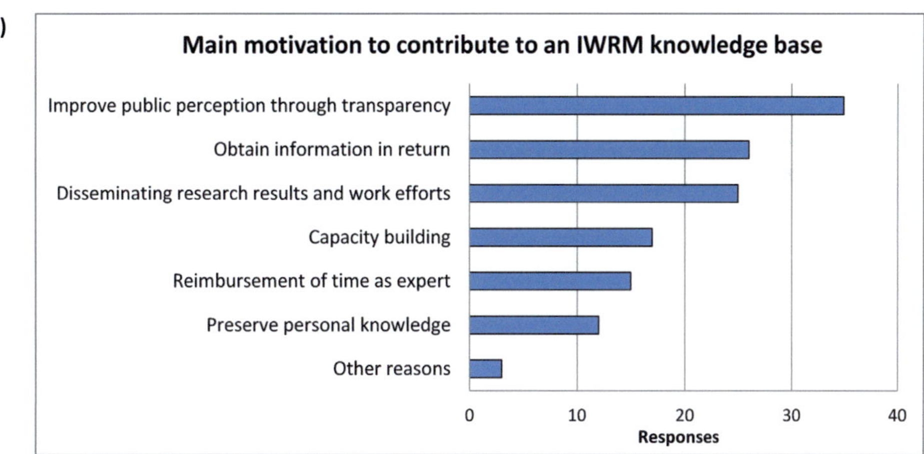

Fig. 4.2: Results from an expert opinion poll about transparency, participation and knowledge management in the Lower Jordan River Valley.

4.3. Requirements Analysis

4.3.1. Terminology

The discussion of knowledge modelling and representation in the following sections will employ a vocabulary which was primarily adapted from the field of ontological engineering. The use of some very common expressions like for example: 'concept', 'class' or 'relationship' might, however, lead to some confusion. Even more so, as this basic vocabulary is not consistently used across different prominent logical knowledge structuring endeavours (e.g. OWL/RDF (Motik et al., 2009), *UMBEL (Bergman & Giasson, 2008), Cyc (Reed & Lenat, 2002), DOLCE (Oberle et al., 2007))*, which then again differ from the debated terminology debate in philosophical ontology, and also have a particularly different understanding when compared to other fields of information and data modelling (for some additional discussion see e.g. Smith et al., 2006).

In comparison to above mentioned formal knowledge structuring specifications the eventual syntax selection as in **Table 4.1**, was considerably condensed in recognition of the reduced logic expressivity of the semantic wiki technology.

Table 4.1: Terminology of the representational units used for knowledge modelling. */**: terms are used as equivalents, depending on whether referring to the general knowledge structure (*) or the implementation into the semantic wiki structure (**).

Term	Understanding	Example	Comparable OWL term
Concept	A mental representation of something constructed by combining all its known characteristics	+ The hydrological concept of a spring + the concept of spring discharge	
Entity	Abstract formalized representation of an individually existing thing about which information is to be stored	+ A spring + a discharge relationship between a spring and a stream	Entity
Individual* Article**	A concrete and named existing individual thing.	+ Hazzir Spring + the average yearly discharge of Hazzir Spring	Individual
Class* Category**	Classification using rules, principles or predicates to determine whether something is an instance of that class.	+ The class of springs + the class of discharge measurements	Class
Relation*/** Property*/**	Hierarchical or non-hierarchical relation between pairs of instances or instances and literals. Relationships between classes are typically hierarchical generalizations.	+ Hazzir Spring is located in Wadi Shueib + Hazzir Spring is a spring + Hazzir Spring has discharge	Object- or Data- Property
Literal	Representation of concrete value of a property	+ 0.96 MCM/a	Literal

4.3.2. Knowledge Management Design Goal

As discussed in the introductory review on knowledge management (cf. Chap. 2.4), the potential approaches on the subject are diverse. With the fundamental objective being the externalization of knowledge and the facilitation of its retrieval (internalization), a knowledge management strategy must be adapted to the given environment and the people that eventually are both: the providing and receiving end of the knowledge.

Based on the experience from the Wadi Shueib IWRM modelling exercise as well as from discussions in the community of a large IWRM project, it was established that 'knowledge in need of managing' essentially relates to collaborative and comprehensive system understanding and to the assumptions that are used to drive planning and decision making within such systems. Knowledge management therefore appears foremost as a means of facilitating the transportation of such system understanding between participants in the process. From this viewpoint, the participants are predominantly sector experts and decision makers, and the knowledge management process basically is the channelling of expert knowledge and decision maker objectives towards the IWRM planning process. Expert knowledge hereby comprises the manifold pieces of information that researchers and practitioners gather, create and employ in their working routine in order to acquire the desired system understanding. These might span from individual calculation or estimation procedures applied to obtain model parameters, over a full model application and up to complete scientific studies. While the latter would at least be likely to progress into the classical channels of publication and reporting, especially the smaller information entities, however valuable, are often never fully disclosed to a wider audience. On the other hand they might well be interesting for other researchers or practitioners working in nearby or comparable areas to be informed about such work, as much as any idea could benefit from being discussed by an interested audience.

Acknowledgement of these requirements led to the broad design decision as illustrated in (**Fig. 4.3**) and the definition of three fundamental roles and classes for the envisaged IWRM knowledge management system:

Fundamental roles:

- **Domain Expert**: Someone who performs analyses on a specialized subject by gathering, combining and inferring information using various sources (literature, models, studies, or his judgment). Creates knowledge in the work process (knowledge creator).

- **IWRM Analyst**: Someone who uses a set of defined goals (policy) and the pool of available knowledge about options and constraints (feasibility, cost-benefit) to perform an IWRM planning and decision process. Also includes the role of **Decision Maker.**

- **Knowledge Engineer (also Administrator)** (not depicted in Fig. 4.3): Someone who maintains the knowledge management system both in technological as well as in structural sense. Evolves and adapts the knowledge management process to the requirements of its participants.

A role describes modes of interaction with the system: an IWRM analyst usually also is a domain expert and any person can become a knowledge engineer if they acquire some insights on the knowledge base structure and the skills for its administration.

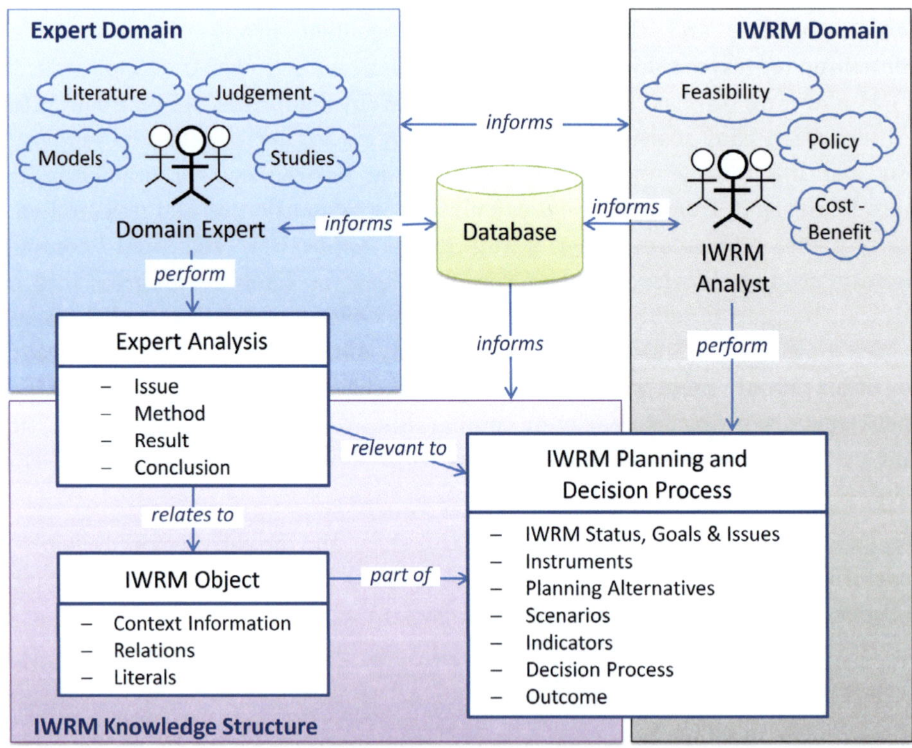

Fig. 4.3: Fundamental concepts and user roles in the IWRM knowledge process.

Fundamental concepts:

- **Expert Analysis**: The expressible product of the analysis process of an expert (e.g. the calculation or estimation of a single value, a literature study, a complex model application or a full scientific study). Related to a distinct entity with significance in the IWRM Process.

- **IWRM Object**: Something that has significance for IWRM, has information about itself and stated relationships to other entities. Can be a tangible entity such as a study area, a spring, a city, a waste water treatment facility, or an intangible entity such as an indicator, a scenario, a literature reference). Can also have a representation in the database.

- **IWRM Process**: IWRM planning and decision process (cf. Chap. 2.3.1) applied to a defined area. Builds on the available information about the associated entities and the given decision constraints and objectives.

Thus, the general approach is to offer an instrument that allows **domain experts** to formulate their **analyses** and structure them by linking to respective interrelated entities (**IWRM-Objects**) which represent the IWRM system. Members of the IWRM domain (**IWRM-Analyst**) on the other hand are enabled to identify and select from the available information to formally document their Decision and Planning processes (**IWRM-Processes**). Formally linking between a database and the entities in the knowledge base (mostly with the IWRM-Objects) establishes an integration of the knowledge management instrument into an overarching information system. The ultimate goal is to evolve a well-structured IWRM knowledge base that provides rich application-level information from experts for experts, IWRM-analysts and decision makers and integrates smoothly into a holistic planning and decision practice.

4.3.3. Challenges of IWRM Knowledge Representation

Basic aspects and general approaches on knowledge representation were already discussed in Chapter 2.4. This section discusses a set of important design considerations for a functional knowledge management setting according to the stated goals and usage environment.

4.3.3.1. *Collaborative Ontology Engineering*

Ideally, an ontology-driven collaborative knowledge management setting would engage its users into an evolutionary ontology development process. Experience shows, however, that ontology development is a very time-intensive task which also requires expertise of dedicated ontology engineers (Lumsden et al., 2011; Simperl, 2009). And even when given users that are experienced in knowledge representation, collaborative methodologies for ontology engineering are still not effectively established (Palma et al., 2011; Vrandecic et al., 2005).

Given the concrete application scenario of this study, an effective collaborative ontology engineering process is very unlikely, as the users are expected to be neither familiar to formal knowledge representation, nor interested in spending time on such conceptual discussions. Thus, the knowledge representation structure has to be widely predefined to allow a functional knowledge sharing and retrieval application. This means that user contributions, that informally extend the ontology, will usually have to be formalized by the dedicated knowledge engineer role.

4.3.3.2.   Level of Formalization

The appropriate level of formalization is strongly dependant on the purpose and eventual application scenarios. Fully formalized ontologies support automated knowledge processes and promise various interesting applications in the IWRM domain (e.g. information retrieval over distributed and independent databases, information exchange between different modelling tools). Full formalization, however, is a complex and slow process that requires commitment of a dedicated user group. Given the broadness of the domain, developing a fully formalized and fine-grained IWRM-ontology would almost compete with huge scale efforts like the Cyc-project (Matuszek et al., 2006) on which already many person-years, maybe centuries, were spent. Furthermore, stronger formalization requires stronger abstraction, and thus separates the knowledge representation layer from the (human) knowledge user layer and requires interfaces that manage the mapping in between.

On the other hand, the anticipated use case has its primary focus on the support of human agents in knowledge acquisition, structuring and dissemination. In the ontology application classification proposed by Jasper & Uschold (1999) this would resemble closest to the "indexing" scenario. In this scenario the principle actors are ontology authors and knowledge workers and the latter employ the ontology mainly to search and identify relevant information from the provided structure. Without the need for inference and strong logic reasoning, full formalization is not necessary for this application scenario. But the knowledge representation has to fulfil the requirements to support content structuring and semantic querying. Following these considerations it appears effective to start with the definition of a semi-formal ontology, which covers essential concepts of the IWRM domain. Necessary formalization will then be added at the implementation phase, to allow for the intended structuring and querying potential.

4.3.3.3. Definition of the IWRM Knowledge Domain

The development of a knowledge representation structure for sharing IWRM knowledge has to deal with a fundamental Chicken-Egg problem: the knowledge domain is ill-defined and very broad. At the beginning, it cannot be generally defined which kind of analyses will eventually be shared, which temporal scale or which units will be used to describe a certain concept, like for example a spring. As mentioned before, it is also very unlikely that domain experts will invest much time and effort in finding consensus on a broad formalization. On the other hand, a system that restricts usage to a small set of well-defined concepts is not likely to be accepted as a useful tool. Thus, the conceptualisation process has to be approached dynamic and to evolve with user activities and that certain concepts and relationships are unidentified in the beginning (Thakker et al., 2011).

4.3.3.4. Knowledge Representation with Semantic MediaWiki

Although meeting some key requirements for collaborative knowledge management, the use of a semantic wiki also brings specific challenges for knowledge representation. As discussed in Chapter 2.4.5, the basic wiki primitive is a page and in most semantic wikis (also the Semantic MediaWiki tool used in this study, cf. Chap. 4.3.7) statements are formulated as binary relations on a page and use that same page as subject. As a result, it is not possible to assert a property to a page and at the same time make an additional statement about that property. This poses no problem as long as stated properties are either unambiguous one-dimensional dates, or the related object is represented by a page of itself. On the page about a spring, for example, the springs altitude above mean sea level can be safely stated as `[[has altitude (amsl)::654m]]`. And it is also straightforward to state `[[is located in::Wadi Shueib]]`, given that Wadi Shueib will be a separate concept described on an own page.

In many cases, however, a property statement requires description of some additional aspects to become unambiguous, while at the same time it is not practical to model it as a separate concept on a separate page. Continuing the above example, the spring page could state `[[has average discharge::1 MCM/month]]`, which, however, is not unambiguous, as it lacks several dimensions: time period of calculation, reference, etc. As a matter of fact, most information units in the IWRM are likely to come with some sort of multidimensionality. While such "n-ary relations" in OWL models handles every multidimensional information element as a separate object-property class, this appears as not practical within a knowledge sharing wiki system: users would need to create new pages for almost every property statement to add formal descriptions.

In an attempt to handle this requirement, Semantic MediaWiki offers a compound datatype (type:record), that contains a short list of values of other datatypes. These "multi-valued properties" cannot be handled as separate data-properties, thus bringing several limitations for querying and formatting.

4.3.4. Use Cases

This section defines seven concrete use cases for the collaborative IWRM knowledge management system. The selection makes no claim of completeness or universality, but is mostly aligned to the course of the parallel development of the Wadi Shueib WEAP IWRM model. Other important influences have been stakeholder and project partner discussions, the general SMART project progress as well as discussions with practitioners from other IWRM initiatives (GLOWA (2009), BMBF-Begleitvorhaben: Vernetzung der Förderaktivität IWRM). The initial use case definitions were used to discover basic design requirements. In the later implementation stage these use cases will be refined into workflows adapted to the given system possibilities.

4.3.4.1. UC-1: Create IWRM-Entity

Scenario Example: Create the Wadi Shueib catchment or another IWRM-Object, e.g. the Hazzir Spring, in the knowledge base to document and share background information and analyses.

Task: Create an IWRM-Entity and add formal relations that describe metadata of the IWRM-Entity, e.g. that the Hazzir Spring is located in the Wadi Shueib catchment, its coordinates, the aquifer it taps, etc.

4.3.4.2. UC-2: Document Expert Analysis

Scenario Example: Expert has collected precipitation data and performed a spatial interpolation to estimate areal precipitation for the Wadi Shueib catchment (i.e. Chap. 3.3.5.1), which he wants to document and share for collaboration.

Task: Create an analysis entity and add formal relations that describe analysis and results and embed the analysis individual into the IWRM-Entity knowledge structure.

4.3.4.3. UC-3: Document IWRM-Process

Scenario Example: IWRM-Analyst starts an IWRM planning study for the Wadi Shueib catchment, following a structured planning approach (cf. Chap. 3.1) which should be documented and shared for collaboration, as well as to invite experts to provide, review and discuss model and scenario assumptions and possible planning options.

Task: Create an IWRM-Process entity and add formal relations that describe metadata, e.g. the study area, the temporal planning horizon, etc. Add entities for the steps of the planning process (status, goals, alternatives, scenarios, indicators, etc.) and link them to analyses in the knowledge base which are used in the IWRM-study.

4.3.4.4. UC-4: Contribute to Existing IWRM-Entities

Scenario Example: User wants to comment on the areal rainfall-interpolation analysis for the Wadi Shueib catchment to state his opinion and add a literature reference.

Task: Edit an existing IWRM-Object/Analysis/Process by adding or changing formal relations and informal content. Original author and other users are informed about the contribution.

4.3.4.5. UC-5: Reuse Knowledge

Scenario Example: User creates an analysis on groundwater recharge of the Wadi Shueib that uses the areal rainfall analysis (i.e. Chap. 3.3.5.4).

Task: Add a generic relation to an analysis to refer to another entity, which then also automatically displays the relation.

4.3.4.6. UC-6: Retrieve Analyses for a Study Area and Subject

Scenario Example: User wants to retrieve a list of all available analyses on water resources in the Wadi Shueib catchment.

Task: Create a query that asks for analyses with a location relation pointing to a specific IWRM-Object and a subject relation pointing to a specific theme.

4.3.4.7. UC-7: Link IWRM-Entity to Representation in Database

Scenario Example: User wants to inform about the available measurement time series of the Hazzir Spring that are stored in the database.

Task: Define an identifier to link from a knowledge base entity to a database object. The knowledge base sends a query to receive information on data-availability and displays the metadata on the respective database object.

4.3.5. Semi-formal IWRM-Ontology

In order to compose an a-priori knowledge representation structure that suits the fundamental concepts (cf. Chap. 4.3.2) as well as the exemplary use cases (cf. Chap. 4.3.4), a semi-formal class structure of IWRM-Entities was developed. The choice of represented concepts was initiated from the set of water system model nodes used in WEAP (cf. Chap. 3.3.7) as well as from the planning process model used in this thesis (cf. 3.1). The focal point was chosen to be the 'catchment' concept, with all spatial relations pointing towards (or from) it. In several iterative modelling steps the initial concept list was refined and organized in hierarchical (*is_a*) and non-hierarchical relations (e.g. *located_in*) until an applicably generic representation of the IWRM modelling and planning domain, as illustrated in **Fig. 4.4**, was found. The ontology fundamentally distinguishes between concrete, tangible entities (e.g. catchments, water demand sites, water resources, measurement stations, etc.), IWRM process entities and expert analyses. In this constellation the tangible entities as well as the IWRM process entities act as basic information anchors to which the dynamic and generic expert analysis class relates to.

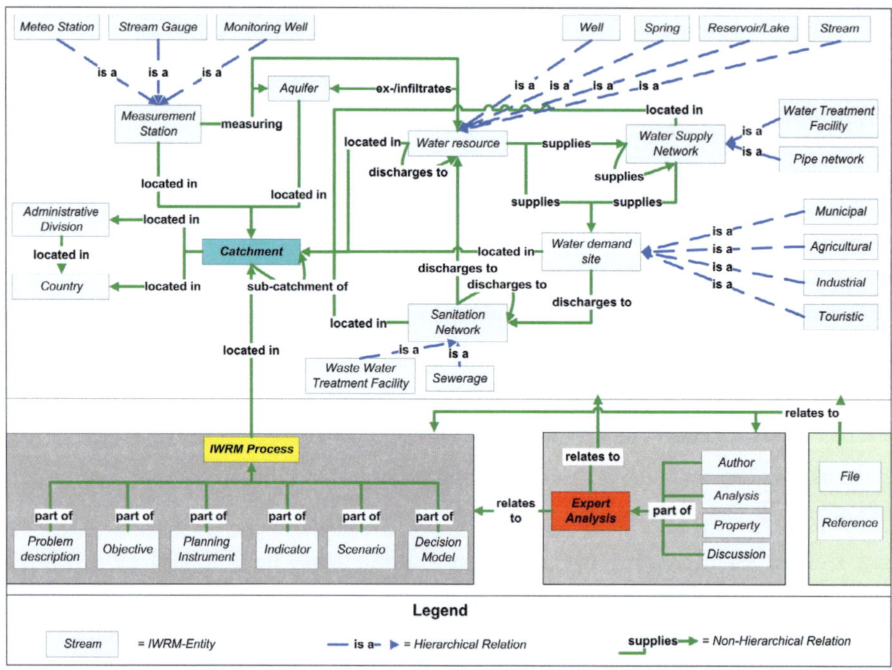

Fig. 4.4: Class-structure of the semi-formal IWRM-ontology.

4.3.6. Requirements

The discussions from Chapters 4.3.2, 4.3.3 and 4.3.4 generate a set of basic design requirements for the IWRM knowledge management platform. Truly, defining the eventual structural, usability and functional requirements has been an iterative process during the work on this thesis. This involved dialogue with project partners as well as experiences from early knowledge representation prototypes, including a DPSIR-like causal network structure (cf. Chap. 2.3.2), an Entity-Attribute-Value data-model as well as a Protégé-ontology modeling exercise. Thus, the requirements listed in **Table 4.2** evolve as much from theoretical considerations as from the experienced practicable feasibility within the scope of the objectives. Eventually, a semantic wiki application was chosen (cf. Chap. 2.4.5), primarily due to the strong collaborative emphasis, the low technological barrier and the general flexibility of the technology.

Table 4.2: Identified requirements for a collaborative knowledge management platform for IWRM planning and decision support.

Knowledge Representation Requirements
IWRM-Entities, Analyses and IWRM-Processes are fully represented.
Multiple representations of knowledge for specific entity and topic are possible.
Formalized relations between all entities are possible.
Representation of n-ary relations.
Formalized relations between entities are represented on both sides (inverse relations).
Support the creation of an analysis-specific formalization vocabulary.

Usability Requirements
Technological barrier for contributors is low.
Users do not need to care about ontology engineering.
Support user in finding the appropriate location in the knowledge base to contribute.
Possibility of generic contribution, if no appropriate location is found.
Possibility to describe entities in a form that is familiar to users, i.e. like writing a short report.
Incomplete formalization of entities is accepted but completion encouraged.
Contributions and edits are immediately represented and communicated.
Spatial information is visualized.
Authorship and contribution is represented.

Functional Requirements
Possibility to add files and external links to all entities.

Support use of relations that are coherent with the ontology.

Edit rights control on IWRM-Entities on contributor side.

Protection mechanism against erroneous edits and/or vandalism.

Interface supports semantic search, facetted browsing and visual information retrieval.

Knowledge Base informs available data from remote data repositories

Knowledge Base is able to interact with external applications

4.3.7. Semantic MediaWiki

Different semantic wiki software products have been under development (Buffa et al., 2008). For this study the "Semantic MediaWiki" (SMW) project (Krötzsch et al., 2006) was chosen, primarily because it is one of the semantic wiki engines that has grown and stabilized during the last years. SMW originated from the AIFB Institute at the University of Karlsruhe as an open source extension to the well-known MediaWiki software and has attracted a large developer and user community today.

SMW extends the MediaWiki syntax with expressions for semantic annotations within the wikitext (cf. Chap. 2.4.5). Annotations are formalized through a mapping to OWL entities. Wiki articles match OWL individuals and the annotations always take their page of occurrence as subject individual to make OWL DL ABox (*assertional box*) statements similar to RDF triples, thus allowing to

- assert a relation (OWL:object-property) to another wiki-page (OWL: individual)
- assert a property (OWL: data-property) and literal of the current page (OWL: individual)
- assert a category (OWL: class) to the current page (OWL: individual)

According to the original developers, *"schematic information (TBox) representable in SMW is kept intentionally shallow [because] [...] it is not intended as a general purpose ontology editor"* (Krötzsch et al., 2006). With its architecture, on the other hand, SMW fosters an approach to implicitly extend the wiki ontology with the annotation of pages. SMW extends the relational MediaWiki database to store the annotation-triples, which means that logic reasoning is partly supported. Free extension modules can be used to employ RDF-Triple Stores that allow for rule modelling and stronger reasoning support.

The following briefly and not comprehensively describes some fundamental structural elements that are used in subsequent sections. Detailed syntax documentation can be found in the user manuals on the MediaWiki (MW, 2012) and Semantic MediaWiki (SMW, 2012) sites and in architectural detail in Krötzsch et al. (2006) and Bao et al. (2008).

Pages are the key structural entities in wikis and also in SMW, as they contain all annotations. Pages are identified by their page-title and are created by calling their URL in or through linking to them from an existing wiki page (e.g. `[[`**New Article Name**`]]`). Basically, it can be differentiated between article pages and non-article pages. Article pages correspond to OWL individuals. It is therefore recommended to see articles as distinct knowledge entities, i.e. a certain spring or a city, that only contain information about themselves, either as implicit context knowledge in text form or as explicit metadata in form of relations and data-properties. Non-article page-types include:

(1) category-, property- and data-type-pages: automatically created on demand and meant to be edited for ontology formalization.
(2) template- and form-pages: created manually as assistant elements.
(3) special-pages: system-created to perform specific functions and usually not meant to be edited.

Properties are automatically created by property-annotations on wiki-pages. For example the article-page of As-Salt containing `[[is located in::Wadi Shueib]]` automatically creates the Property:*Is_located_in* with the default *data-type range:page* and instantiates an individual thereof on the As-Salt page with value:Wadi Shueib. Changing the default data-type (page) to another type, e.g. a number, can be done on the non-article page of the property by annotating it with `[[has type::number]]`. Thereby, SMW allows definition of OWL: object properties (page-types) and OWL: data properties (other types).

Categories are automatically created by a category-annotation on a page to add an OWL-SubClassOf-relation to a page. Annotating for example the As-Salt page with `[[Category::Municipality]]`, classifies the article-individual as a municipality. Adding the category annotation to a category page creates a hierarchical relation. SMW does not restrict the amount of categories a page belongs to, but understanding categories as OWL:Classes recommends using a single (or few) category-classification.

Semantic Internal Objects (SIO) are an extension to the SMW core that provides an additional parser functions to define objects and associated properties from within a wiki page. Internal Objects are annotated on a page by

```
{{#set_internal:object_to_page_property|property1=value1…}}
```

and handled as page-like objects (pages without an article representation), thus, allowing their properties to be queried separately. SIOs can be used to represent n-ary relations in a wiki environment, without the need to create new pages for every multidimensional information element.

Inline queries are inserted into a page and allow querying annotations in the wiki ontology. The query syntax is SMW-specific and uses the pattern

```
{{#ask: [[condition 1]] [[...]] [[condition n]]
|?printout parameter 1|?...|?printout parameter n
|format parameter 1|...| format parameter n}}
```

The condition argument are used for selection and can include category selectors, property selectors, comparators (for property selections), OR disjunctions and property chains. While the condition arguments return a list of pages, the printout parameters state which page properties are to be fetched and displayed in the query results. Format parameters define the display format results (e.g. as list, table, map, etc.).

Templates are pages of the template-namespace that are embedded into other pages, when called by

```
{{name_of_the_template_page|optional_parameter_1|...|n.
```

While the template page contains wiki-text, as well as placeholders for the parameters as *optional_parameter_1}}*.

Thereby, templates generally enable structuring, formatting and reusing of wiki content and are already very popular in non-semantic MediaWiki applications (e.g. Wikipedia). In the SMW community, the use of templates is also perceived as very advantageous, as they provide an interface between wiki editors and the wiki ontology. In this respect, using template parameters to handle semantic annotations basically saves users from the necessity to search the ontology structure for the correct properties before contributing formal content.

Semantic Forms are an extension to SMW that allows using form-based input masks for wiki editing and querying. Forms are constructed as pages in the Form-namespace where the input fields and their behaviour are defined and mapped to one or multiple templates. The definition of input field behaviour reflects certain aspects of the rdfs:domain and rdfs:range properties in OWL which can be used to support as well as control user input. Actual semantic annotations are still defined within the templates, thus, making the semantic forms another interface layer on top of templates.

4.4.   Implementation

With the DROPEDIA platform, a prototype of a collaborative knowledge management platform for IWRM planning and decision support has been implemented in the course of this thesis. After an initial closed development phase, the platform was opened with online access to the SMART project partners (full access) and the public (read only) under http://129.13.109.100/~dropedia/ (**Fig. 4.5**). The following sections provide an overview of essential architecture elements that were implemented to meet the identified requirements (cf. Chap. 4.3.6).

4.4.1.   Hardware and Software

Based on the classical web-based wiki-concept, the Dropedia-platform is hosted as a client-server-model. At the time of publication of this thesis, the webserver was running on a self-hosted AMD Athlon 64 3000+ (1.8 GHz) machine. Furthermore, Dropedia is completely based on Open Source Software distributed under the GNU General Public License in versions 2.0, 3.0 or under compatible licenses approved by the Open Source Initiative (OSI). The wiki itself is running on a *MediaWiki* installation with a number of extensions that are mostly related to the *Semantic MediaWiki* project. A detailed list of the employed software packages is found in Annex B.

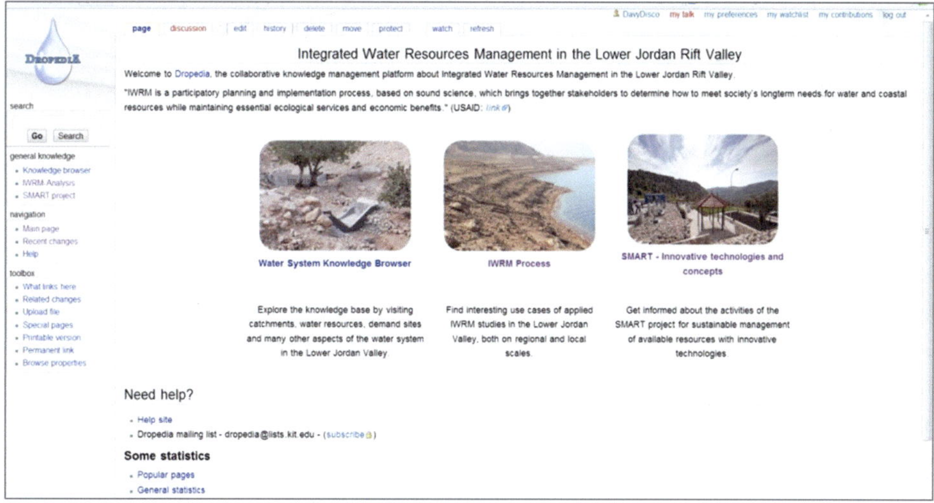

Fig. 4.5: DROPEDIA welcome screen.

4.4.2. Architecture

4.4.2.1. Dropedia-Ontology

An empty wiki can be understood as a plain sketchbook, thus it may evolve in a multitude of different ways but it gives no initial guidance to its user. In order to assist the users in the development of a structured knowledge repository the semi-formal IWRM-ontology (cf. Chap. 4.3.5) was fully implemented as organizing base structure for DROPEDIA. With regard to aforementioned matters of knowledge representation in SMW (cf. Chap. 4.3.3.4), the formalization has to deal with some specific modelling requirements. For specification of the rdfs- and swivt-vocabulary please see RDFS (2010) and SWIVT (2010).

As a first step, all a-priori identified IWRM-Entity classes are represented as categories. Categories themselves can be annotated as members of a category, thus, allowing a *rdfs:subClassOf* hierarchy that can be used for structuring and querying. Abundant categorization can, however, also reduce clarity. Therefore, the category hierarchy is kept shallow in the DROPEDIA-Ontology, constituting a maximum of three levels with the top-classes: IWRM-Object, IWRM-Process and Analysis.

For every IWRM-Entity a set of properties was identified that provide a basic formal description of the subject. Formal description of a complex concept, such as a city or a spring, can quickly lead to a high amount of perceivably necessary predicates, especially when considering available knowledge about specific individuals. The latter typically results in the formulation of relations that an entity can have, for example that a spring can have a protection zone, but at the same time distracts from the fundamental attributes that all individuals of the class share. In order to control the amount of necessary properties, the descriptive predicates were iteratively condensed to find least common denominators of property definitions. Eventual properties include a set of descriptions that can be reused for different concepts (e.g. spatial descriptions like *is_located_in, has_coordinates, has_area,* or general descriptions like *has_reference, has_file*). Category-specific properties are only added for very common and stable attributes, (e.g. *has_aquifer_tapped* for springs and wells). All specific and dynamic attributes, on the other hand, are considered as instances of the Analysis category, so to allow arbitrary specification as well as multiple attribute instances.

Property definition includes the various built in types, as well as a set of customized data-types, (e.g. area, flow, water type) that allow additional control on input (*swivt:allowedValue*) and representation (*swivt:mainUnit* and *swivt:conversionFactor*).

Since the SMW-model, as OWL/RDF in general, does not support propagation of properties from categories to individuals, the assertion of the metadata properties was realized by assigning templates to each category which handle annotations when an instance of a category is created.

SMW also lacks native support of property ranges (*rdfs:range*) which can significantly help users in ontology-coherent knowledge formulation. As an informal workaround, property ranges were implemented via input data types in Semantic Forms input field definitions (with "`values from= x`", where x is a specific category, property or namespace).

A subset of the eventual DROPEDIA-Ontology is illustrated in **Fig. 4.6**.

The combination of several requirements posed a challenge on the DROPEDIA-Ontology:

1. Users must be able to represent multiple views on one property. Therefore, and because information cannot practically be atomized to unambiguous one-dimensionality in a wiki environment, users must be able to represent n-ary relations in DROPEDIA (as discussed in Chap. 4.3.3.4 SMW lacks native support of true n-ary relations).
2. An a-priori knowledge model cannot provide formalisms for all probable representation requirements. Therefore, users should be able to extend the basic vocabulary with own terms.
3. Users should not have to bother with ontology issues and are thus strongly supported by predefined form inputs (cf. Chap. 4.4.2.2).

Take for example the representation of the population number of a municipality, having different values for two different censuses, as well as for different scales (urban or district). As every number can have relevance for the documentation of an IWRM process, it is not useful to simply decide on one number and drop the other information. Furthermore, another user does not employ population numbers, but the number of housing units in his IWRM model, for which no pre-defined property exists.

Separating knowledge representation into IWRM-Object entities and related Analysis entities basically solves the first of above issues. In doing so, an IWRM-Object (e.g. a municipality) can relate to any number of views (e.g. have several related population analyses).

The creation of a generic formalization mechanism with the use of the Semantic Internal Objects (SIO) extension tackles the second and third issue. Each Analysis individual can contain an unlimited number of SIOs which represent user-defined n-ary properties.

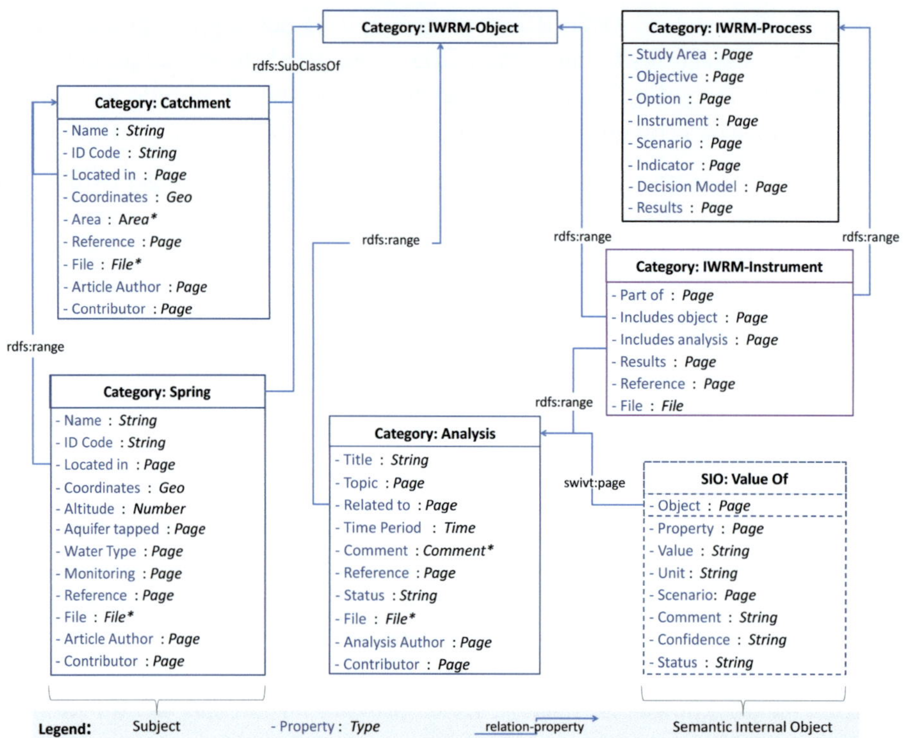

Fig. 4.6: Subset of the DROPEDIA-Ontology illustrating some basic IWRM subjects and their respective relations and types (custom data-types are marked with *).

4.4.2.2.   *User Interface and Interaction Architecture*

DROPEDIA is primarily intended to interact with human agents. In order to facilitate ontology-coherent contribution without the hindrance of first searching and understanding the DROPEDIA-Ontology, the platform was set up to make strong use of templates and semantic forms for user input and content presentation. **Fig. 4.7** illustrates the setup of the interface architecture from a user perspective. All IWRM-Entity categories have assigned specific input forms that can be used to add and edit articles. Forms always consist of input fields for the metadata properties of the respective category, as well as a free-text input possibility. For most fields input is voluntary, so as to enable contribution even when not all facts are known. Saving an edited form inserts a template call into the wiki text of the article page with the input as template parameters. The transcluded templates fully handle page layout, semantic annotations as well as ask- and SPARQL-queries. Besides the encouraged use of forms and templates, directly editing a pages wiki-text is also kept as possibility, in case the controlled environment is not sufficient.

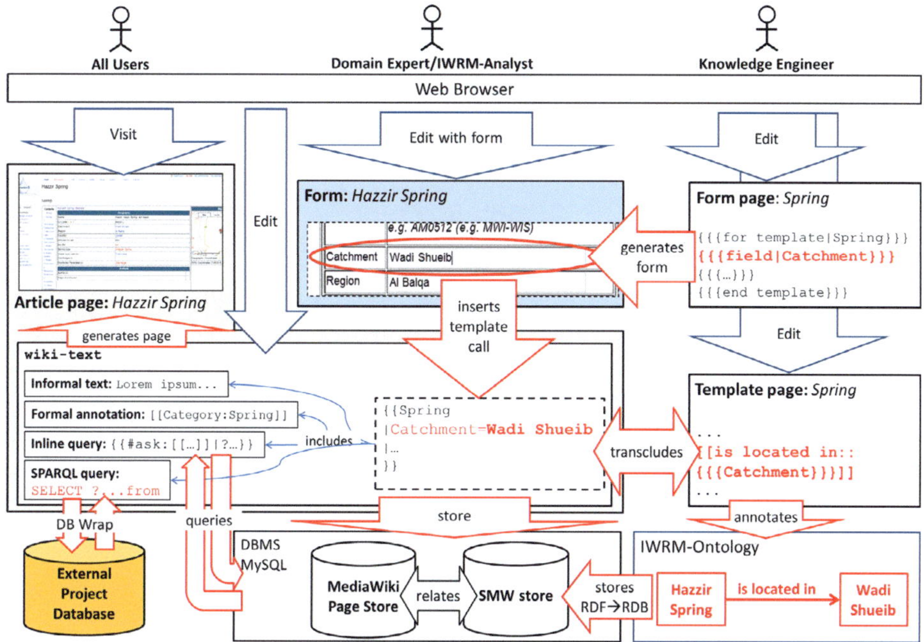

Fig. 4.7: User interface architecture of DROPEDIA. All users can browse pages. Domain Experts and IWRM Analysts can use forms to edit the information on the pages. The Knowledge Engineer ensures semantics and functionality by providing the forms and templates.

SMW and the Semantic Forms extensions provide special meta-forms to create new forms and templates. However, as illustrated in **Fig. 4.8**, the complexity of nested and multiple template-calls, and the layout structuring integrated therein, makes this option impractical. Thus, maintenance and extension of forms and templates are understood as dedicated task for the Knowledge Engineer role. In order to control the required maintenance, forms and templates employ a uniform structure and are generally designed for reuse between categories. Still, at the time of publication, a total of about 50 forms and almost 100 templates of various sizes are in active use on DROPEDIA.

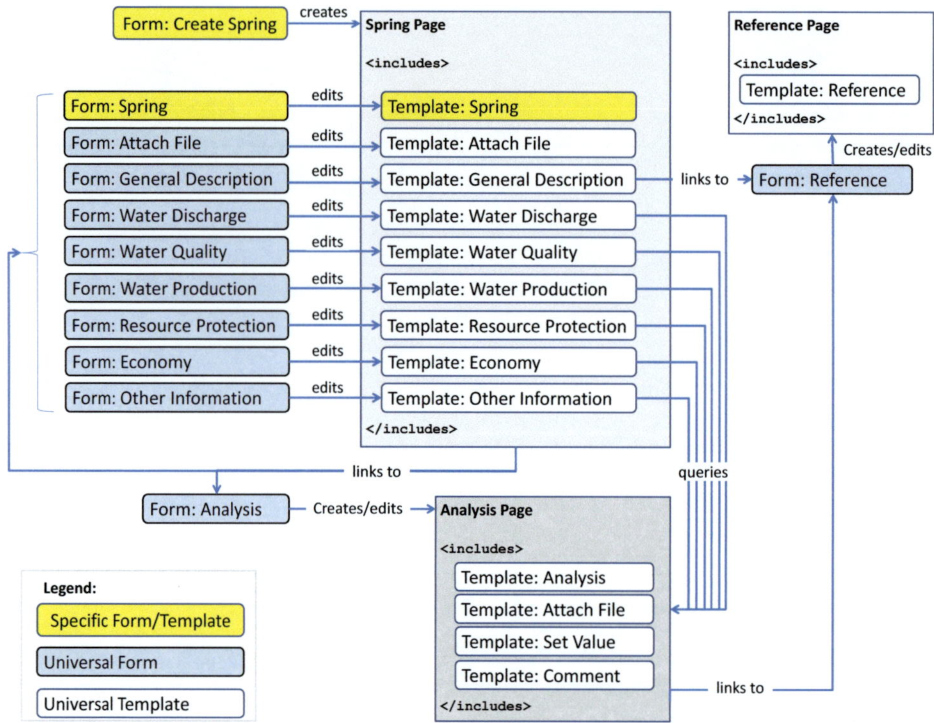

Fig. 4.8: Subset of the form and template structure of DROPEDIA. Specific forms and templates are available for every IWRM-Object category. Universals are reused across categories. Forms and templates for the IWRM-Process category are not depicted.

4.4.2.3.   *Connection to the External Project Database*

As a wiki-based knowledge management platform, DROPEDIA does not compete with relational database management and is neither intended, nor fit, for the efficient handling of large alphanumerical or spatial datasets. On the other hand, it is very useful to let the knowledge base inform users about the availability of data on a specific IWRM-Object in remote data repositories. In this sense the wiki platform can act as an integrative frontend that lets users combine information

For this purpose, a query module that allows representation of SMART Database content in DROPEDIA pages was developed. Implementation of this module was primarily done by Benedikt Kämpgen (Kaempgen et al., in preparation). The general architecture is illustrated in **Fig. 4.9**. A fundamental problem in the design was the restrictive firewall-policy of the SMART Database, as well as the fact that it uses different identifiers as DROPEDIA. Thus, the implemented query module receives the data by HTTP request from a Linked Data Wrapper that wraps daily generated database views from the HYDROSMART XML/HTTP interface. The wrapped data is filled into an online triple store (Qcrumb.com), thus allowing semantic queries from DROPEDIA to be issued with a SPARK query syntax.

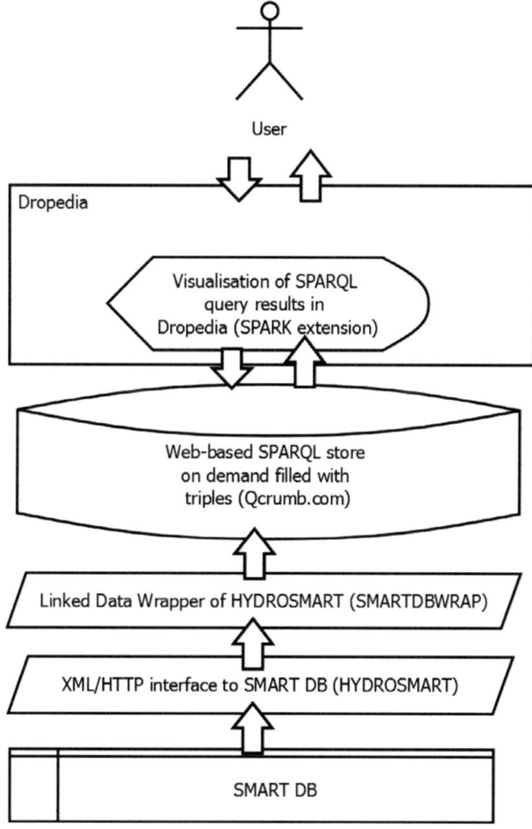

Fig. 4.9: Architecture of DROPEDIA/SMART-DB connection (Kaempgen et al., in preparation).

4.4.3. Use Case Workflows

This section briefly illustrates how the previously defined use cases (UC; cf. Chap. 4.3.4) were realized as workflows in DROPEDIA. For this section, the description follows the illustration in **Fig. 4.10**, although, there are several other possible paths to navigate and interact with the platform to achieve the tasks defined in the use cases.

Users may start on the welcome page or via the permanent left-side navigation and search panel (1) to open to the DROPEDIA Knowledge Browser page (2) which offers an overview of the knowledge base content. Since the overview is automatically created with inline queries at every page load, it always presents an up to date view on the content. The query results are organized as IWRM-Objects in sections that correspond to the DROPEDIA -Ontology categories and are displayed in table as well as in map format.

To add a new IWRM-Entity to the knowledge base (UC-1: cf. Chap. 4.3.4.1), every section offers an "Add *[IWRM-Object]*"-link (3) that opens the respective input form (4) with fields for the categories metadata-properties. Various helpful features of the Semantic Forms extension are realized, including: auto-completion of input according to the defined property ranges, file uploads, map-navigation for geographic locations as well as geocoding of addresses. Saving the form (5) creates the article page for the IWRM-Object (6), displaying the defined metadata properties, attached files and a series of sections that represent primary topics of the subject (e.g. a spring page shows sections for water discharge, water quality, e.g.).

Every section can be edited by any (logged in) user by following the displayed "Add/Edit *[Section]*" links that lead to a corresponding form input (7). Generally, every IWRM-Object, Analysis or IWRM-Process page offer respective links, thus allowing successive updating and commenting of the knowledge base content (UC-4: cf. Chap. 4.3.4.4).

Article pages also present a graph-view (8) on available data in the SMART Database, created by an embedded SPARQL query at every page load (UC-7: cf. Chap. 4.3.4.7).

To create an analysis and add it to the IWRM-Object of the current article, users can follow the "Add *[Section]* Analysis"-link (9) to open the Analysis-form (10) that provides input fields for Analysis metadata properties, file uploads and free text, as well as multiple instance form fields to add an unlimited number of user-defined properties that are stored as SIOs   (UC-2: cf. Chap. 4.3.4.2). Furthermore, the *related_to* property can hold a list of IWRM-Object- or Analysis pages to create relations and reuse information between knowledge elements in DROPEDIA (UC-5: cf. Chap. 4.3.4.5). Saving the Analysis form creates an Analysis-page that displays the input and queries the stored SIOs. The latter are also displayed in a query on the article page of the related IWRM-Object (11) (UC-5: cf. Chap. 4.3.4.6). The workflow of documenting an IWRM-process (UC-3: cf. Chap. 4.3.4.3) is not illustrated in **Fig. 4.10**, but it basically follows the same scheme as the creation of an IWRM-Object.

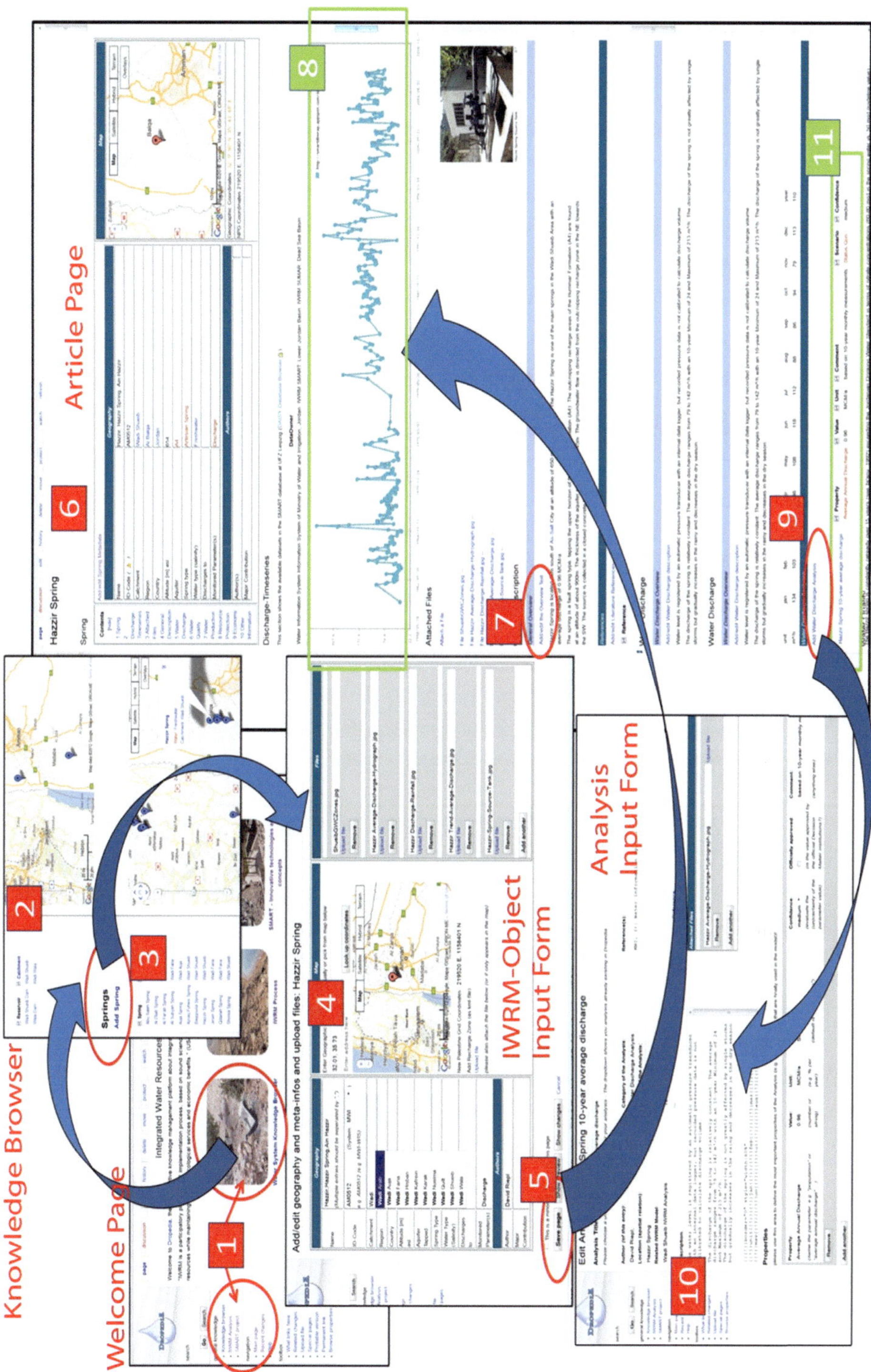

Fig. 4.10: DROPEDIA workflow for IWRM-Object and Analysis contribution.

4.5. Evaluation

The following section evaluates to which extent the previously identified IWRM knowledge management requirements (cf. Chap. 4.3.6) could be met in DROPEDIA and which requirements could not or have not (yet) been implemented. At the time of publication, DROPEDIA has already been opened to project partners, but has so far not achieved extensive usage within the SMART project. Thus, the presented evaluation has to rely in large parts on a preliminary and informal assessment of the knowledge platform potential.

4.5.1. Knowledge Representation Requirements

IWRM-Entities, Analyses and IWRM-Processes are (fully) represented.	yes

The statements of the preliminary semi-formal IWRM-Ontology are represented in the DROPEDIA-Ontology and all concepts are described by a set of meta-data properties. Full representation, however, also implies completeness, which appeared as not feasible in the broadness of the IWRM domain. In this regard, the DROPEDIA-Ontology is meant to evolve with its use.

Multiple representations of knowledge for specific entity and topic are possible.	yes
Representation of n-ary relations.	(yes)
Support the creation of an analysis-specific formalization vocabulary.	yes

The separation of IWRM-Object entities and related Analysis entities provides a working solution for the representation of multiple views and n-ary relations.

Knowledge representation is not necessarily fully formalized, as the level of aggregation or segmentation of analyses as well as the optional formulation of properties with SIOs, is basically left to user choice. Yet, the three layered model of: IWRM-Object $\leftarrow$ *is_related_to* $\leftarrow$ Analysis $\rightarrow$ *has_value* $\rightarrow$ SIO, does not represent n-ary properties in a strict sense and requires nested queries for retrieval. Also users might not intuitively understand this threefold separation.

Formalized relations between all entities are possible.	(yes)

Basically, by editing a pages wiki-text any relation can be defined. This was only partly realized in the input forms, basically for the Analysis *is_related_to* and the IWRM-Process *includes* relations. Generic relations between IWRM-Object might be another useful addition to the DROPEDIA-Ontology.

Formalized relations between entities are represented on both sides (inverse properties).	(no)

In newer versions of SMW it is possible to query for the inverse direction of a property. However, this does not assign the property to the targeted object.

4.5.2. Usability Requirements

Technological barrier for contributors is low.	yes
Users don't need to worry about ontology engineering.	(yes)

The wiki environment generally set the technological barrier low. Users only need a standard web-browser to access DROPEDIA and, owing to Wikipedia, are usually well adapted to the typical wiki-style navigation. The structural formalism introduces an abstract complexity that was heavily overbuilt by the use of forms and templates. On the other hand, several interesting functionalities are still reserved to proficient users with some understanding of the SMW markup and the DROPEDIA-Ontology. Foremost to mention are the possibility to create own queries or to extent the ontology.

Support user in finding the appropriate location in the knowledge base to contribute.	yes
Possibility of generic contribution, if no appropriate location is found.	yes

Several structural elements were implemented for this purpose. Especially the DROPEDIA-Knowledge Browser is designed to provide quick and structured overview of the knowledge base content and allow direct access to contribution. But as mentioned before, the necessary separation between the fundamental concepts is not intuitive and is likely to result in some need of introduction for new users. All input forms also provide free text fields for generic and informal contributions. Furthermore, any contributor has the possibility to create unstructured ("normal") wiki pages.

Possibility to describe entities in a form familiar to users, i.e. like writing a short report.	(yes)

Pre-setting of a certain structural requirement is unavoidable for formalization. And it disregards the advantages of the semantic structuring to just copy/paste reporting text into DROPEDIA pages. Still, due to abovementioned informal input possibilities, as well as the generic setup of the Analysis category, the level of formalization is to the greatest possible extent left to user choice.

Incomplete formalization of entities is accepted but completion encouraged.	yes

Only few metadata properties critical for structuring are set to mandatory input.

Contributions and edits are immediately represented and communicated.	(yes)

New contributions are by default directly represented on the edited pages. Furthermore, several native MediaWiki-tools like "History", "Recent changes", or "Related changes" support tracking wiki changes. Interesting changes might, however, be obscured, as these functions record all changes, regardless of the significance. The Knowledge Browser provides an automatically updated structured overview, as do the immediate query representations of new Analysis relations on IWRM-Object pages. Further interesting and also readily available improvements could for example include a "Newest Article" feature on the main page as it is known from other wikis, or the possibility to mark specific pages of interest in order to receive automatic email change notifications.

Spatial information is visualized.	yes

With the use of the Maps and Semantic Maps extensions, the representation of point-coordinates and kml-files is possible on the base maps of several web mapping services (googlemaps, openlayers, osm, yahoomaps). Inline queries can have map representation as result format.

Authorship and contribution is represented.	yes

To emphasize authorship and contribution beyond the native and relatively anonymous wiki standard, the specific metadata-properties *has_author* and *has_contributor* were introduced. Secondly, editing rights are restricted to registered users.

4.5.3. Functional Requirements

Possibility to add files and external links to all entities.	yes

File uploads and external links are basic MediaWiki functionalities that were consistently employed in the input forms.

Support use of relations that are coherent with the ontology.	yes

General ontology-coherence is realized through the use of templates for annotation handling and semantic forms for user input. A possibly useful addition thereto, could be to offer of an ontology-browser interface that allows more proficient users to conveniently explore categories, properties and instances. The native possibility, to list these entities on their corresponding Namespace-overviews is too unrelated to help understanding of the structuring.

Protection mechanism against erroneous edits and/or vandalism.	yes
Edit rights control on IWRM-Entities on contributor side.	no

Wikis are fundamentally built on the principle of open editing. Nonetheless, the "Revision History"-function of wiki-pages provide a capable protection mechanism against erroneous or unwanted edits. Another possibility also discussed for DROPEDIA, was to give authors of articles the control on edit rights. The Access Control List MediaWiki extension that introduces such functionality was tested in the prototype phase, but eventually dropped due to compatibility problems.

Interface supports semantic search, facetted browsing and visual information retrieval.	(yes)

Facetted navigation is generally supported by the article-templates and the Knowledge Browser, which both support map display and query-generated structured content. Semantic search is also possible, but still lacks satisfactory support for non-proficient users, as it requires understanding of SMW markup and the ontology structure.

Knowledge Base informs available data from remote data repositories	(yes)
Knowledge Base is able to interact with external applications	(no)

An informative connection to the SMART-Database was realized, which allows to query a daily database-view and represent the results in DROPEDIA-Pages. The modular approach is suitable to be extended to other distributed data-repositories. External application interaction is also basically supported through the SMW functionality that allows the import and export of RDF statements. This, however, was not tested until the publication of this thesis.

4.6. Summary

This chapter has presented a conceptual framework and a prototype implementation for the structured and collaborative documentation and sharing of planning and decision relevant knowledge in the IWRM process.

As a first step, a generic model of knowledge management requirements was deduced from the Wadi Shueib modelling exercise (cf. Chap. 3.3), and iteratively adapted to the requests of the community of scientists and water sector professionals in the SMART project. In order to approach the challenge of selecting and formally representing relevant knowledge elements from the highly heterogeneous IWRM domain, a set of concrete use cases and a semi-formal IWRM-Ontology was introduced. And out of this, a list of specific structural, functional and usability requirements was generated.

With the development of the DROPEDIA prototype, it has been shown how the majority of the defined requirements can be realized on the basis of the Semantic MediaWiki-technology. Here, especially the transformation and subsequent specification from the semi-formal IWRM-Ontology to the formal DROPEDIA-Ontology was a challenging and important development step. In order to support contributors in ontology-consistent knowledge formulation and retrieval, a user-centred interface model was developed which is based on complex and nested templates and semantic forms. Thereby, DROPEDIA achieves a combination of the collaborative, generic and ease-of-use nature of a wiki environment and the strengths of a semantically enriched repository for efficient knowledge structuring and retrieval.

4.7. Conclusions

4.7.1. Introducing Semantic Knowledge Management to IWRM

The demand for improved communication and transparency between participants in the IWRM process is regularly pointed out in the literature. It was also ascertained and exemplified in the course of evolving the Wadi Shueib IWRM planning model (cf. Chap. 3.3). At the same time, contemporary knowledge management with semantic web techniques has a high potential to provide efficient support to these ends, as successful applications in other disciplines have frequently shown (cf. Chap. 4.1). The IWRM domain, however, still lacks exploitation of this potential.

In this regard, the presented work is the **first consequent development of a semantic knowledge management framework for collaborative planning and decision processes in Integrated Water Resources Management.**

As it was thoroughly discussed throughout Chapter 2.4, there cannot be a generic knowledge management strategy, applicable to any given objective, environment or user group. Therefore, the focus was placed on what was perceived as one of the most urgently required aspects: the collaborative documentation and sharing of planning and decision relevant system understanding during the IWRM process. **Using the Wadi Shueib example as central blueprint, it could be demonstrated, how such processes can be reasonably formalized and efficiently managed in a semantic structure that is strongly directed towards knowledge sharing.**

It is expected that comparable activities can greatly benefit from the use of the presented framework or in general from the application of semantic knowledge management techniques. The incentive on the side of researchers and practitioners lies primarily in the prospect to receive a direct and fast feedback from a widened community of interest as well as from their stakeholders already during the course of their work. Stakeholder could use this channel to stay constantly informed about on-going research as well as to take influence by expressing their direct requirements. Moreover, better documentation of models and modelling processes usually resides high on the wish lists of practitioners that adopt models from research initiatives for their own work. And another very essential incentive for enhanced documentation is the considerable loss of skilled personal ("brain drain"), under which especially developing countries are constantly suffering.

4.7.2. Expressive IWRM-Knowledge Representation with Ontologies

Aspects of water sector knowledge have been previously represented in ontologies. Earlier studies, however, focused mostly on strict and rule-based representations of small sub-domains, with the objective of enabling expert system-like functionalities for user-guidance (e.g. Ceccaroni et al., 2004; Scholten, 2008; Volk et al., 2007).

Representing the broadness of the IWRM domain, while at the same time finding the appropriate formalization level between generality and usability has been an ambitious challenge. **In this regard, the semi-formal IWRM-Ontology (cf. Chap. 4.3.5), as well as the deduced DROPEDIA-Ontology (cf. Chap. 4.4.2.1), are unique developments** that eventually achieved a satisfactory expressive potential for the applied use cases (cf. Chap. 4.4.3). Nonetheless, the ontologies are profoundly meant to evolve with their use.

Although, the collaborative nature of the wiki environment has a potential to foster such evolution (Schaffert et al., 2005), the level of complexity that has already been reached is likely to keep most users from considering active ontology extension. Besides, the IWRM domain is not a discipline with a very strong focus on strict formalization of its knowledge assets.

The more rational approach is to assign this task completely to the knowledge engineer role. It would include the regular assessment of usage patterns for emerging representation requirements and the consequential structure translation. Together with the necessary maintenance and extension of the template and form-driven user interaction model, this emphasizes the importance of a strong and dedicated knowledge engineer role for running a collaborative knowledge management platform in the complexity of the IWRM domain. **A clearly structured uniform template- and form architecture as proposed in Chap. 4.4.2.2 significantly helps with the required maintenance and necessary extensions.**

4.7.3. Semantic MediaWiki for Collaborative IWRM Knowledge Management

Options in the landscape of knowledge management instruments are plentiful (cf. Chap. 2.4.3). The original starting point for the exploration of semantic knowledge management options was the search for an effective structuring technique, applicable to the complex and heterogeneous information requirements of the IWRM domain. Recognition of the essential necessity of a strong collaborative element, in addition to structural functionalities, led to the eventual choice of the semantic wiki approach.

Although only few alternative semantic wiki engines have been tested, it is safe to say that the use of Semantic MediaWiki brings several essential advantages: the software has continuously grown and stabilized during the last years, it has already attracted a large and active developer and user community today, and it directly extends the very well-known MediaWiki engine.

The DROPEDIA prototype demonstrates the potential of a semantic wiki framework to provide solutions for various critical knowledge management requirements (cf. Chap. 4.5). In the IWRM context, it was experienced that especially a low technological-barrier and a general easy usability are essential for contributors to adapt to a new system. Thus, in order to handle the structural formalism DROPEDIA consequently employs semantic forms and templates to build a straightforward user-centred interface-architecture. In conjunction with the structural setup of the DROPEDIA-Ontology and the use of Semantic Internal Objects, this also provides users control on the formalization level of their contributions, while at the same time they don't have to be concerned about the tedious and complex task of ontology engineering.

Overall, it was not evaluated, if the semantic wiki way is the best or most suitable approach to realize a collaborative knowledge management platform for IWRM planning and decision support. However, for the most part, **the initially experienced limitations of the wiki-nature of knowledge representations could be overcome with the presented structural setup.** Eventually, it was demonstrated how DROPEDIA is able to store, display and edit information in many types of rich content (e.g. text, data values, references, images, maps, files, videos), and how **structured queries could be realized for internal as well as external information from remote data repositories.**

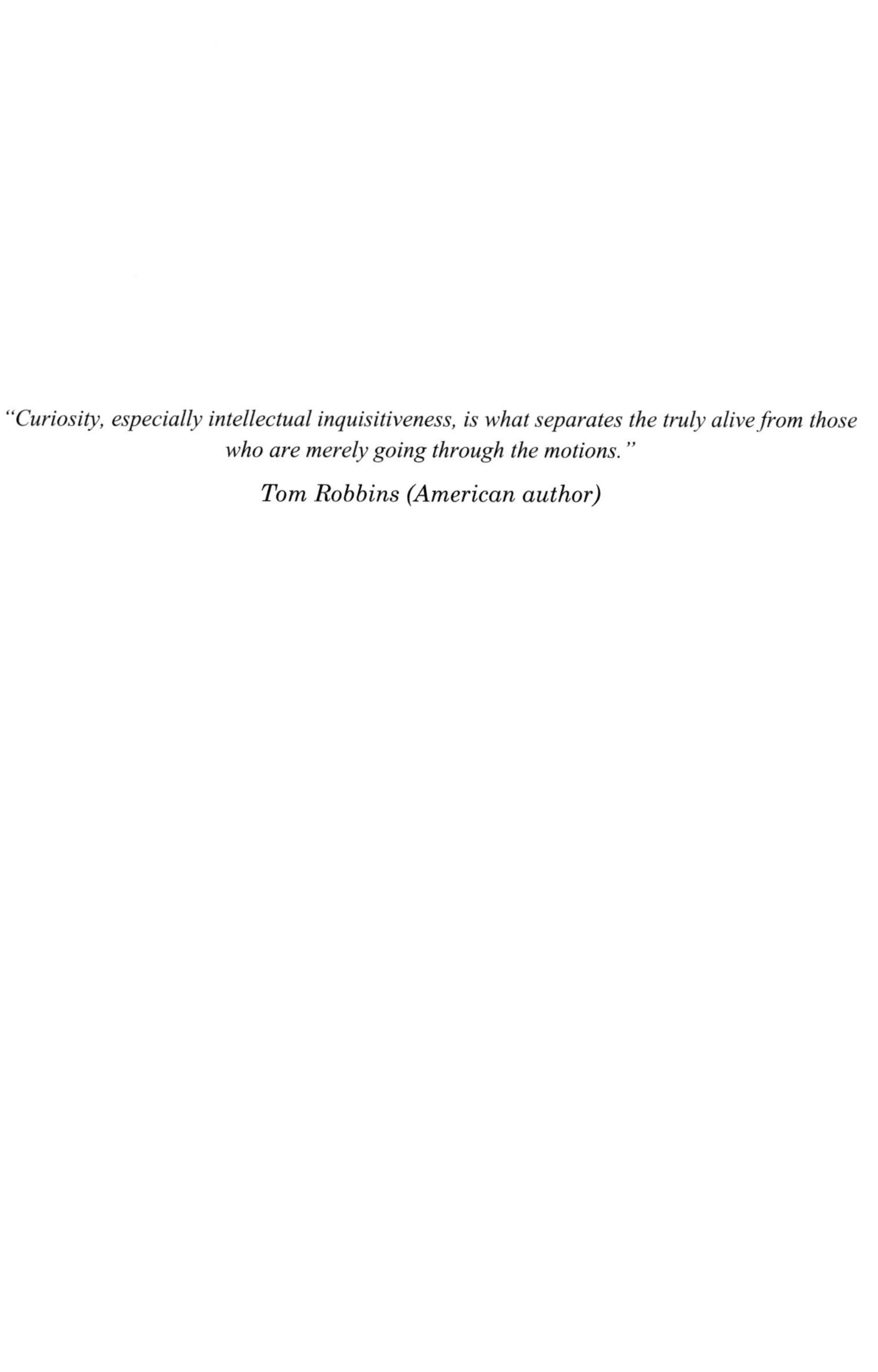

"Curiosity, especially intellectual inquisitiveness, is what separates the truly alive from those who are merely going through the motions."

Tom Robbins (American author)

5.   Research Summary and Outlook

Previous sections have already provided in-depth discussions, summaries and conclusions on the IWRM modelling and scenario planning study (cf. Chap. 3.5) and the collaborative knowledge management approach and platform (cf. Chap. 4.6).

Thus, this section will concentrate on a brief summary with regard to the major contributions and give the author's thoughts on interesting future research potentials in relation to the presented work.

This thesis has three major contributions

1. The comprehensive analysis and holistic integration of a large set of available data and information from various sources significantly enhances the understanding of the Wadi Shueib water resources system, provides the basis for future integrated water resources management analyses, and also highlights critical gaps in the current state-of-knowledge and the potential for future research initiatives.
2. A demonstration of an IWRM modelling and scenario planning framework on a sub-basin-scale, that consequently combines national strategic development objectives with local planning options, and provides a useful indicator toolkit for evaluating scenario performance in terms of IWRM principles.
3. The introduction of a contemporary semantic knowledge management approach to the IWRM domain, with an ontology prototype for collaborative IWRM planning and decision support knowledge sharing, and a semantic wiki architecture that also supports non-technicians in the documentation, semantic structuring and representation of such knowledge.

5.1. Outlook and Recommendations

Based on the set of available data (cf. Chap. 3.3.3), it was demonstrated that it is possible to establish the operational Wadi Shueib IWRM WEAP model (cf. Chap. 3.3.7) for water balancing and scenario planning. Extending the model to further study areas on a similar scale and detail-level would of course be highly interesting. Data acquisition and quality control, however, was an extensive task. And in this regard, it has to be kept in mind that the Wadi Shueib catchment is regarded as very well studied in Jordan. Using the Wadi Shueib model, as presented here, as a blueprint, could prove useful to early identify the information requirements of a comparable IWRM model and help increasing the data acquisition efficiency.

Regarding the presented Wadi Shueib model itself, it has been mentioned (cf. Chap. 3.3.6), that some important compartments and flow paths of the water balance are still subject to high uncertainty. Especially groundwater related aspects bear interesting potential for future detail studies. And here, regarding the complex karst lithology of the catchment area, tracing techniques with a focus on unintentional recharge appear a feasible and rational approach. Overall, water quality aspects have not been in the scope of the presented Wadi Shueib WEAP model, but are currently addressed by another working group within the SMART project.

The scenario planning approach was found especially useful in combination with the detailed analysis of the national objectives and the local planning possibilities. Therefore, extending this analysis to other study areas could bring immediate further insights into gaps and possible bridges between IWRM theory and practice. In this regard, the extension of the presented IWRM performance indicator set could result in a truly comprehensive and generic indicator list applicable to future IWRM studies.

The semantic knowledge management approach presented in this study has been a novel way for an IWRM initiative. In this regard, it had constantly to deal with a wicked chicken-egg problem: Ideally, a functional knowledge management would have been available from the beginning of the project – but to know which and how IWRM knowledge has to be managed could only be learned during the project. In this respect, future IWRM initiatives are urgently advised to specifically address this task. And not only as a side activity assigned to domain experts or the overall project management, but as a dedicated effort for fostering communication and collaboration.

The presented knowledge management approach has also focused primarily on knowledge sharing. A multitude of emerging applications demonstrated in other disciplines, suggest a considerable potential to extend the presented framework to further knowledge management tasks, like knowledge acquisition or knowledge application. One example could be to investigate the possibility to include the growing open data repositories (e.g. Linked Open Data (LOD, 2012)) into queries conducted from within the knowledge base.

DROPEDIA also bears potential for interesting development potential in terms of structural features. First of all, it could be useful to review and the current caching strategy of DROPEDIA. With the Alternative PHP Cache (APC, 2012) and the Squid cache (SQUID, 2012), two supportive caching modules have already implemented which appear to improve parsing response times, but there has not yet been a systematic validation.

Due to stability reasons, DROPEDIA has been kept running on an SMW version, that has already been outdated at the time of publication of this thesis. It could be very interesting to thoroughly investigate the added functionality of new core and extension versions, especially in respect of newly added querying and result presentation possibilities. A possible use case could be the dynamic graph representation of semantic content, e.g. in the form of a water balance scheme that dynamically represents its elements according to their data values.

Eventually, a more thorough evaluation of the user interface and the use cases could provide valuable insights into future enhancement potential. For example by applying small scale usability tests as presented in Lange (2011), or, after a longer period as an evaluation of usage patterns and discussion culture.

5.2. Critical Remark: Sharing Knowledge – Experiences from DROPEDIA

At the time of publication of this thesis, the discussed DROPEDIA platform had been opened to contributions by SMART project partners and stakeholders for about one year. Thus, it appears time for a preliminary review on the adaption.

Concept, design and functionality generally perceived approval from the project community as well as from the international IWRM community, when presented at international symposia (Riepl et al., 2011). And sector experts signalled interest and willingness to invest time into a wiki-based knowledge platform (cf. Chap. 4.2).

However, any knowledge management system is dependent on the input from a motivated user community, but adaption of DROPEDIA has not yet reached the anticipated level. There might be several reasons for the reluctance of the targeted audience. For a start, a brief review of other online communities reveals related challenges that are often called the "1% rule" or the "Participation Inequality". The basic hypothesis is that in most online communities user participation more or less follows a 1-9-90 distribution, with 1 % regular and 9 % occasional contributors as well as 90 % that only consume. In his blog, Nielsen (2006) collected some interesting figures on the topic, including some of the largest online communities (e.g. Wikipedia, Amazon), that all more or less apply to such a participation divide. This also relates to some part to the experience of the author, as the most often asked question after a DROPEDIA-presentation was rather: "what can I find there?" instead of "how can I contribute/collaborate?".

Another important aspect of reluctant contribution might be the difficulty to design a good integration into the day-to-day workflow of the targeted audience, due to their heterogeneity. In this regard, contributing might be perceived as additional work. And the potential incentive of collaboration can only be encouraged by citing successful endeavours from other disciplines, but eventually has to become a personal experience. A possible way out of this dilemma would require the creation of additional incentives, as for example, to explicitly channel reporting duties into a knowledge sharing platform.

And a third essential reason is most likely linked to the fear of losing control over one's personal knowledge assets and in the worst case having others copy or exploit the personal work. This fear, actually has long been driven in the quasi-economic competitive principle of scientific research (Knorr-Cetina, 1982). At least here, a possible way out may just be on the brink of emergence, as new types of scientists and practitioners begin to embrace the idea of opening their notebooks (WP, 2012c) and governments (WP, 2012b), opening their data (WP, 2012a) and knowledge (WP, 2012d) and eventually opening their minds.

The End. Stop. ...or?

6.   References

Abecker, A. (2004). Business-Process Oriented Knowledge Management: Concepts, Methods, and Tools. *PhD-Thesis*, Universität Karlsruhe (TH), Karlsruhe. Available from http://digbib.ubka.uni-karlsruhe.de/volltexte/1000003269

Abu-Ajamieh, M. M. (1998). *Water Quality Improvement and Conservation Project: Assessment of Brackish Groundwater in Jordan*: USAID.

Abu-Jaber, N. S., Aloosy, A. S. E., & Ali, A. J. (1997). Determination of aquifer susceptibility to pollution using statistical analysis. *Environmental Geology, 31*(1), 94-106. http://dx.doi.org/10.1007/s002540050168

Ackoff, R. L. (1989). From data to wisdom. *Journal of Applied Systems Analysis, 16*, 3-9

Acreman, M. (2005). Linking science and decision-making: features and experience from environmental river flow setting. *Environmental Modelling & Software, 20*(2), 99-109. http://dx.doi.org/10.1016/j.envsoft.2003.08.019

Al-Kloub, B., & Abu-Taleb, M. F. (1998). Application of Multicriteria Decision Aid to Rank the Jordan-Yarmouk Basin Co-riparians According to the Helsinki and ILC Rules. *Water International, 23*(3), 164-173. http://dx.doi.org/10.1080/02508069808686763

Al-Kuisi, M. (1998). *Effects of irrigation water with special regards to biocides on soils and groundwater in the Jordan Valley area, Jordan*. Münster: Verein der Geologie-Studenten.

Al-Omari, A., Al-Quraan, S., Al-Salihi, A., & Abdulla, F. (2009). A Water Management Support System for Amman Zarqa Basin in Jordan. *Water Resources Management, 23*(15), 3165-3189. http://dx.doi.org/10.1007/s11269-009-9428-z

Alavi, M., & Leidner, D. E. (2001). Review: Knowledge Management and Knowledge Management Systems: Conceptual Foundations and Research Issues. *MIS Quarterly, 25*(1), 107-136. http://www.jstor.org/stable/3250961

Albani, R., Soer, G., & Tarawneh, T. (2011). Evaluation of Instituting Water Demand Management *Project Evaluation Reports*. I. P. Jordan. Amman: USAID.

Alcamo, J., Doll, P., Henrichs, T., Kaspar, F., Lehner, B., et al. (2003). Global estimates of water withdrawals and availability under current and future "business-as-usual" conditions. *Hydrological Sciences Journal-Journal Des Sciences Hydrologiques, 48*(3), 339-348. http://dx.doi.org/10.1623/hysj.48.3.339.45278

Alexa, The Web Information Company. (2012). Alexa Page Ranking. Retrieved 2012-01-08, from http://www.alexa.com/siteinfo/wikipedia.org

Alfaro, P., Riepl, D., Seder, N., Subah, A., Wolf, L., et al. (2012). *Coupling of a Water Evaluation and Planning System Model with a Groundwater Model in Wadi Shueib - Jordan.* - Poster presented at the IWRM 2012, Karlsruhe, November 21-22 2012.

Alfarra, A., Kemp-Benedict, E., Hötzl, H., Sader, N., & Sonneveld, B. (2011). A Framework for Wastewater Reuse in Jordan: Utilizing a Modified Wastewater Reuse Index. *Water Resources Management, 25*(4), 1153-1167. http://dx.doi.org/10.1007/s11269-010-9768-8

Alkhoury, W. (2011). Hydrological modelling in the meso scale semiarid region of Wadi Kafrein / Jordan - The use of innovative techniques under data scarcity. *PhD-Thesis*, Georg-August-Universität zu Göttingen, Göttingen. Available from http://webdoc.sub.gwdg.de/diss/2011/alkhoury/

Allen, R. G., Pereira, L. S., Raes, D., & Smith, M. (1998). Crop evapotranspiration: guidelines for computing crop water requirements, FAO irrigation and drainage paperpp. XXVI, 300 S). Available from http://www.fao.org/docrep/X0490E/x0490e00.htm

Amatya, D., Skaggs, R., & Gregory, J. (1995). Comparison of Methods for Estimating REF-ET. *Journal of Irrigation and Drainage Engineering, 121*(6), 427-435. http://dx.doi.org/10.1061/(ASCE)0733-9437(1995)121:6(427)

Amery, H. A. (2002). Water wars in the Middle East: a looming threat. *Geographical Journal, 168*(4), 313-323. http://dx.doi.org/10.1111/j.0016-7398.2002.00058.x

Anderson, R. L., Steinitz, C., Arias Rojo, H. M., & Bassett, S. (2003). *Alternative futures for changing landscapes: the Upper San Pedro River Basin in Arizona and Sonora.* Washington; London: Island Press.

Andrews, I. J. (1992). *Permian, triassic and jurassic lithostratigraphy in the subsurface of Jordan.* Amman: Surbsurface Geology Division.

APC, PHP Manual. (2012). Alternative PHP Cache Documentation. Retrieved 2012-02-12, from http://php.net/manual/en/book.apc.php

Arnold, J. G., & Allen, P. M. (1999). Automated methods for estimating baseflow and ground water recharge from streamflow records. *Journal of the American Water Resources Association, 35*(2), 411-424. http://dx.doi.org/10.1111/j.1752-1688.1999.tb03599.x

Asian Development Bank. (1999). *Handbook for the economic analysis of water supply projects.* Manila: Asian Development Bank - Economics and Development Resource Center.

Baker, J. P., Hulse, D. W., Gregory, S. V., White, D., Van Sickle, J., et al. (2004). Alternative futures for the Willamette River Basin, Oregon. *Ecological Applications, 14*(2), 313-324. http://dx.doi.org/10.1890/02-5011

Banat, K. M., Howari, F. M., & Al-Hamad, A. A. (2005). Heavy metals in urban soils of central Jordan: Should we worry about their environmental risks? *Environmental Research, 97*(3), 258-273. http://dx.doi.org/10.1016/j.envres.2004.07.002

Bao, J., Ding, L., & Hendler, J. (2008). Knowledge Representation and Query in Semantic MediaWiki: A Formal Study *Technical Report*. Tetherless World Constellation (RPI) Retrieved from http://tw.rpi.edu/wiki/TW-2008-42

Barlow, M., & Clarke, T. (2002). *Blue gold: the battle against corporate theft of the world's water.* London: Earthscan.

Batty, M., & Xie, Y. (1994). Modelling inside GIS (Part 1): Model structures, exploratory spatial data analysis and aggregation. *International Journal of Geographical Information Systems, 8*(3), 291-307. http://dx.doi.org/10.1080/02693799408902001

Bayer, H. J., Hoetzl, H., Jado, A. R., Roscher, B., & Voggenreiter, W. (1988). Sedimentary and Structural Evolution of the Northwest Arabian Red-Sea Margin. *Tectonophysics, 153*(1-4), 137-151. http://dx.doi.org/10.1016/0040-1951(88)90011-X

Bdour, A. M. (2012). Perspectives on a Strategic Jordanian Water Project: The Red Sea to Dead Sea Water Conveyance Construction. *Journal of Emerging Trends in Eng. & App. Sciences, 3*(1)

Becker, A., & Hattermann, F. (2005). Model-supported Participatory Planning for Integrated River Basin Management (Deliverable No. D3/11-13) *Deliverables*. H.-C. W. 3: Harmoni-CA-Peoject.

Becker, K. (2000). Hydrological and Hydrogeological Investigations along Wadi Shueib, Jordan. *Unpublished Diploma-Thesis*, University of Karlsruhe, Karlsruhe.

Bellinger, G., Castro, D., & Mills, A. (2004). Data, Information, Knowledge, and Wisdom. *Systems Thnking*. Retrieved from http://www.systems-thinking.org/dikw/dikw.htm

Ben Hussein, M. (2010). Fuheis residents demand relocation of cement factory. *Jordan Times*, April 18th 2010.

Bender, F. (1974). *Geology of Jordan*. Berlin: Gebrüder Borntraeger.

Bergman, M., & Giasson, F. (2008). UMBEL Ontology *Technical Documentation*. Retrieved from http://techwiki.umbel.org/index.php/UMBEL_Specification

Berners-Lee, T., Hendler, J., & Lassila, O. (2001). The Semantic Web - A new form of Web content that is meaningful to computers will unleash a revolution of new possibilities. *Scientific American, 284*(5), 34-+

Besleme, K., & Mullin, M. (1997). Community indicators and healthy communities. *National Civic Review, 86*(1), 43-52. http://dx.doi.org/10.1002/ncr.4100860107

Beven, K. (2006). A manifesto for the equifinality thesis. *Journal of Hydrology, 320*(1-2), 18-36. http://dx.doi.org/10.1016/j.jhydrol.2005.07.007

BGR, & MWI. (2010). Delineation of Groundwater Protection Zones for the Springs in Wadi Shuayb *Groundwater Resources Management*: Federal Ministry for Economic Cooperation and Development.

Binney, D. (2001). The knowledge management spectrum – understanding the KM landscape. *Journal of knowledge management, 5*(1), 33-42. http://dx.doi.org/10.1108/13673270110384383

Biswas, A. K. (2004). From Mar del Plata to Kyoto: an analysis of global water policy dialogue. *Global Environmental Change, 14, Supplement*(0), 81-88. http://dx.doi.org/10.1016/j.gloenvcha.2003.11.003

Biswas, A. K., Tortajada, C., & Izquierdo-Avino, R. (2009). Water management in 2020 and beyond. In A. K. Biswas (Eds.), Water resources development and management Available from http://dx.doi.org/10.1007/978-3-540-89346-2

Black, E. (2010). Water and society in Jordan and Israel today: an introductory overview. *Philosophical Transactions of the Royal Society a-Mathematical*

Physical and Engineering Sciences, 368(1931), 5111-5116. http://dx.doi.org/10.1098/rsta.2010.0217

Black, E., Hoskins, B., Slingo, J., & Brayshaw, D. (2011). The Present Day Climate in the Middle East. In S. Mithen & E. Black (Eds.), *Water, Life and Civilisation: Climate, Environment and Society in the Jordan Valley.* Cambridge: Cambridge University Press.

Black, E. W. (2008). Wikipedia and academic peer review: Wikipedia as a recognised medium for scholarly publication? *Online Information Review, 32*(1), 73-88. http://dx.doi.org/10.1108/14684520810865994

Blaney, H. F., & Criddle, W. D. (1950). *Determining water requirements in irrigated areas from climatological and irrigation data.* Washington, D.C.: U.S. Soil Conservation Service.

Blanning, R. W. (1979). The functions of a decision support system. *Information & Management, 2*(3), 87-93. http://dx.doi.org/10.1016/0378-7206(79)90039-9

Borgstedt, A., & Subah, A. (2008). *Groundwater Protection and Sanitation - Practical experiences in Jordan.* Paper presented at the International Symposium on Coupling Sustainable Sanitation & Groundwater Protection.

Bossel, H. (1996). Deriving indicators of sustainable development. *Environmental Modeling and Assessment, 1*(4), 193-218. http://dx.doi.org/10.1007/bf01872150

Bossel, H. (1999). *Indicators for sustainable development: theory, method, applications: a report to the Balaton Group.* Winnipeg: IISD International Institute for Sustainable Development.

Brachman, R. J. (1985). On the Epistemological Status of Semantic Networks. In R. J. Brachman & H. J. Levesque (Eds.), *Readings in Knowledge Representation* (pp. 191-215). Los Altos, CA: Kaufmann.

Bradfield, R., Wright, G., Burt, G., Cairns, G., & van Der Heijden, K. (2005). The origins and evolution of scenario techniques in long range business planning. *Futures, 37*(8), 795-812. http://dx.doi.org/10.1016/j.futures.2005.01.003

Bradley, J.-C., Owens, K., & Williams, A. (2008). Chemistry Crowdsourcing and Open Notebook Science. *Nature Precedings (online).* http://dx.doi.org/10.1038/npre.2008.1505.1

Brans, J.-P., & Mareschal, B. (1994). The PROMCALC & GAIA decision support system for multicriteria decision aid. *Decision Support Systems, 12*(4–5), 297-310. http://dx.doi.org/10.1016/0167-9236(94)90048-5

Buckley, J. J. (1985). Fuzzy hierarchical analysis. *Fuzzy Sets and Systems, 17*(3), 233-247. http://dx.doi.org/10.1016/0165-0114(85)90090-9

Buffa, M., Gandon, F., Ereteo, G., Sander, P., & Faron, C. (2008). SweetWiki: A semantic wiki. *Web Semantics: Science, Services and Agents on the World Wide Web, 6*(1), 84-97. http://dx.doi.org/10.1016/j.websem.2007.11.003

Butler, D. (2005). Joint efforts. *Nature, 438*(7068), 548-549. http://dx.doi.org/10.1038/438548a

Camargo, M. B. P., & Hubbard, K. G. (1999). Spatial and temporal variability of daily weather variables in sub-humid and semi-arid areas of the United States high plains. *Agricultural and Forest Meteorology, 93*(2), 141-148. http://dx.doi.org/10.1016/S0168-1923(98)00122-1

Cardwell, H. E., Cole, R. A., Cartwright, L. A., & Martin, L. A. (2006). Integrated Water Resources Management: Definitions and Conceptual Musings. *Journal of Contemporary Water Research & Education, 135*(1), 8-18. http://dx.doi.org/10.1111/j.1936-704X.2006.mp135001002.x

Cariaso, M., & Lennon, G. (2012). SNPedia: a wiki supporting personal genome annotation, interpretation and analysis. *Nucleic Acids Research, 40*(D1), D1308-D1312. http://dx.doi.org/10.1093/Nar/Gkr798

Cash, D., Clark, W., Alcock, F., Dickson, N., Eckley, N., et al. (2002). Salience, Credibility, Legitimacy and Boundaries: Linking Research, Assessment and Decision Making. *Harvard University: KSG Faculty Research Working Papers Series, RWP02*(046). http://dx.doi.org/10.2139/ssrn.372280

CAWMA. (2007). *Water for food water for life: a comprehensive assessment of water management in agriculture.* London: Earthscan.

Cazier, D. J., & Hawkins, R. H. (1984). *Regional application of the curve number method.* Proceedings of the Water Today and Tomorrow: Irrigation and Drainage Division Special Conference, New York.

Ceccaroni, L., Cortés, U., & Sànchez-Marrè, M. (2004). OntoWEDSS: augmenting environmental decision-support systems with ontologies. *Environmental Modelling & Software, 19*(9), 785-797

Chermack, T. J., Lynham, S. A., & Ruona, W. E. A. (2001). A review of scenario planning literature. *Futures Research Quarterly 17*(2), 25

Chiew, F. H. S., Kamaladasa, N. N., Malano, H. M., & McMahon, T. A. (1995). Penman-Monteith, FAO-24 reference crop evapotranspiration and class-A pan data in Australia. *Agricultural Water Management, 28*(1), 9-21. http://dx.doi.org/10.1016/0378-3774(95)01172-f

Choi, B., & Lee, H. (2003). An empirical investigation of KM styles and their effect on corporate performance. *Information & Management, 40*(5), 403-417. http://dx.doi.org/10.1016/s0378-7206(02)00060-5

Clerc, V., de Vries, A. E., & Lago , A. P. (2010). *Using wikis to support architectural knowledge management in global software development.* Proceedings of the 2010 ICSE Workshop on Sharing and Reusing Architectural Knowledge, Cape Town, South Africa.

Cole, M. (2009). Using Wiki technology to support student engagement: Lessons from the trenches. *Computers & Education, 52*(1), 141-146. http://dx.doi.org/10.1016/j.compedu.2008.07.003

Darari, F., & Manurung, R. (2011). *LinkedLab: A Linked Data platform for Research Communities.* Paper presented at the International Conference on Advanced Computer Science and Information System (ICACSIS).

De Roure, D., Goble, C., Aleksejevs, S., Bechhofer, S., Bhagat, J., et al. (2010). Towards open science: the myExperiment approach. *Concurrency and Computation: Practice and Experience, 22*(17), 2335-2353. http://dx.doi.org/10.1002/cpe.1601

Dingman, S. L. (2008). *Physical Hydrology* (2nd ed. ed.). Long Grove, Ill.: Waveland.

Dirks, K. N., Hay, J. E., Stow, C. D., & Harris, D. (1998). High-resolution studies of rainfall on Norfolk Island: Part II: Interpolation of rainfall data. *Journal of Hydrology, 208*(3–4), 187-193. http://dx.doi.org/10.1016/s0022-1694(98)00155-3

Ditzel, P. (2008). Transformation of water organizations. From administrative hierarchy beyond marketization. The case of Amman Water, Jordan. *MSc Thesis*, Zeppelin University, Friedrichshafen. Available from http://www.zu.de/deutsch/forschung_forschungsprojekte/MA-Award_FT08_PMG_PhilippDitzel.pdf

DoS. (1994). Jordan National Population and Housing Census. Amman: Department of Statistics of the Hashemite Kingdom of Jordan.

DoS. (2004). Jordan National Population and Housing Census. Amman: Department of Statistics of the Hashemite Kingdom of Jordan.

Drucker, P. F. (1995). The information executives truly need. *Harvard Business Review, 73*(54). http://dx.doi.org/10.1225/95104

Eckhardt, K. (2005). How to construct recursive digital filters for baseflow separation. *Hydrological Processes, 19*(2), 507-515. http://dx.doi.org/10.1002/Hyp.5675

EEA. (2005). EEA core set of indicators *Technical report*. European Environmental Agency. Luxembourg: European Environmental Agency.

Ekins, S., Hupcey, M. A. Z., & Williams, A. J. (2011). *Collaborative computational technologies for biomedical research*. Hoboken, N.J.: Wiley.

El-Naqa, A., & Al-Shayeb, A. (2009). Groundwater Protection and Management Strategy in Jordan. *Water Resources Management, 23*(12), 2379-2394. http://dx.doi.org/10.1007/s11269-008-9386-x

Esty, D. C., Levy, M., Srebotnjak, T., & de Sherbinin, A. (2005). Environmental sustainability index: benchmarking national environmental stewardship. New Haven: Yale Center for Environmental Law and Policy.

EXACT. (1998). *Overview of Middle East Water Resources*.: US Geological Survey.

Fahey, L., & Prusak, L. (1998). The eleven deadliest sins of knowledge management. *California Management Review, 40*(3), 265-+

Fahey, L., & Randall, R. M. (1998). *Learning from the future: competitive foresight scenarios*. New York: Wiley.

Falkenmark, M. (1986). Fresh-Water - Time for a Modified Approach. *Ambio, 15*(4), 192-200

Falkenmark, M. (1989). The Massive Water Scarcity Now Threatening Africa - Why Isnt It Being Addressed. *Ambio, 18*(2), 112-118. http://www.jstor.org/stable/4313541

Fane, S., Robinson, J., & White, S. (2003). The use of levelised cost in comparing supply and demand side options. *Water Supply, 3*(3), 7

Faucher, J. P., Everett, A. M., & Lawson, R. (2008). Reconstituting knowledge management. *Journal of knowledge management, 12*(3), 3-16. http://dx.doi.org/10.1108/13673270810875822

Fedra, K., Weigkricht, E., & Winkelbauer, L. (1992). Decision Support and Information Systems for Regional Development Planning *Problems of Economic Transition: Regional Development in Central and Eastern Europe*. T. Vasko. Avebury, Aldershot, UK.

Firestone, J. M., & McElroy, M. W. (2004). Organizational learning and knowledge management: the relationship. *Learning Organization, The, 11*(2), 177-184. http://dx.doi.org/10.1108/09696470410521628

Fischhendler, I., & Heikkila, T. (2010). Does Integrated Water Resources Management Support Institutional Change? The Case of Water Policy Reform in Israel. *Ecology and Society, 15*(1)

Fleischbein, K., Wilcke, W., Valarezo, C., Zech, W., & Knoblich, K. (2006). Water budgets of three small catchments under montane forest in Ecuador: experimental and modelling approach. *Hydrological Processes, 20*(12), 2491-2507. http://dx.doi.org/10.1002/hyp.6212

Flexer, A., Guttman, J., Haddad, M., Hötzl, H., & Rosenthal, E. (2009). The Water of the Jordan Valley: Scarcity and Deterioration of Groundwater and Its Impact on the Regional Development. In H. Hötzl, P. Möller & E. Rosenthal (Eds.). Berlin, Heidelberg: Springer.

Freund, R., Garfunkel, Z., Zak, I., Goldberg, M., Weissbrod, T., et al. (1970). The Shear along the Dead Sea Rift [and Discussion]. *Philosophical Transactions of the Royal Society of London. Series A, Mathematical and Physical Sciences, 267*(1181), 107-130. http://dx.doi.org/10.1098/rsta.1970.0027

Frické, M. (2009). The knowledge pyramid: a critique of the DIKW hierarchy. *Journal of Information Science, 35*(2), 131-142. http://dx.doi.org/10.1177/0165551508094050

Galaz, V. (2007). Water governance, resilience and global environmental change - a reassessment of integrated water resources management. *Water Science and Technology, 56*(4), 1-9

Garfi, M., & Ferrer-Marti, L. (2011). Decision-making criteria and indicators for water and sanitation projects in developing countries. *Water Science and Technology, 64*(1), 83-101. http://dx.doi.org/10.2166/Wst.2011.543

Garshol, L. M. (2004). Metadata? Thesauri? Taxonomies? Topic Maps! Making Sense of it all. *Journal of Information Science, 30*(4), 378-391. http://dx.doi.org/10.1177/0165551504045856

Gausemeier, J., Fink, A., & Schlake, O. (1998). Scenario Management: An Approach to Develop Future Potentials. *Technological Forecasting and Social Change, 59*(2), 111-130. http://dx.doi.org/10.1016/s0040-1625(97)00166-2

Genesereth, M. R., & Nilsson, N. J. (1988). *Logical foundations of artificial intelligence* (Reprint. with corr. ed.). Palo Alto: Kaufmann.

Geoffrion, A. M. (1983). Can Ms/or Evolve Fast Enough. *Interfaces, 13*(1), 10-25

Geyh, M. A., Rimawi, O., & Udluft, P. (1985). *Environmental isotope study in the Hamad region Natural water groups and their origin of the shallow aquifers complex in Azraq-depression, Jordan.* Stuttgart: Schweizerbart.

Giannini, V., & Giupponi, C. (2011). Integration by identification of indicators. *Adv. Sci. Res., 7*, 5. http://dx.doi.org/10.5194/asr-7-55-2011

Giles, J. (2005). Internet encyclopaedias go head to head. *Nature, 438*(7070), 900-901. http://dx.doi.org/10.1038/438900a

Giupponi, C. (2007). Decision Support Systems for implementing the European Water Framework Directive: The MULINO approach. *Environmental Modelling & Software, 22*(2), 248-258. http://dx.doi.org/10.1016/j.envsoft.2005.07.024

Giupponi, C., & Sgobbi, A. (2008). Models and Decisions Support Systems for Participatory Decision Making in Integrated Water Resource Management.

In P. Koundouri (Ed.), *Coping with Water Deficiency* (Vol. 48, pp. 165-186): Springer Netherlands.

Gleick, P. H. (2003). Global Freshwater Resources: Soft-Path Solutions for the 21st Century. *Science, 302*(5650), 1524-1528. http://dx.doi.org/10.1126/science.1089967

GLOWA. (2009). An integrated approach to sustainable management of water ressources under global change: GLOWA Jordan River; Phase II - final report. Eberhard-Karls-Universität Tübingen - Fakultät für Biologie - Botanisches Institut. Tübingen [u.a.]. Retrieved from http://download.glowa-jordan-river.com/

Godet, M. (2000). The Art of Scenarios and Strategic Planning: Tools and Pitfalls. *Technological Forecasting and Social Change, 65*(1), 3-22. http://dx.doi.org/10.1016/s0040-1625(99)00120-1

Goodwin, P., & Wright, G. (2001). Enhancing Strategy Evaluation in Scenario Planning: a Role for Decision Analysis. *Journal of Management Studies, 38*(1), 1-16. http://dx.doi.org/10.1111/1467-6486.00225

Goovaerts, P. (1997). *Geostatistics for natural resources evaluation.* New York: Oxford Univ. Press.

Grace, T. P. L. (2009). Wikis as a knowledge management tool. *Journal of knowledge management, 13*(4), 64-74. http://dx.doi.org/10.1108/13673270910971833

Gregory, K. J., & Walling, D. E. (1976). *Drainage basin form and process: a geomorphological approach* (repr ed.). London: Arnold.

Grey, D., & Sadoff, C. W. (2007). Sink or Swim? Water security for growth and development. *Water Policy, 9*(6), 545-571. http://dx.doi.org/10.2166/Wp.2007.021

Grimmeisen, F., Zemann, M., Sawarieh, A., Wolf, L., & Goldscheider, N. (2012). *Grundwasserschutz in urban geprägten Karstgebieten mit semi-aridem Klima - Fallstudie Hazzir-Quelle, Wadi Shueib, Jordanien.* - Poster presented at the FH-DGG Tagung 2012, Dresden 2012.

Groves, D. G., Yates, D., & Tebaldi, C. (2008). Developing and applying uncertain global climate change projections for regional water management planning. *Water Resources Research, 44*(12). http://dx.doi.org/10.1029/2008wr006964

Gruber, T. R. (1993). A Translation Approach to Portable Ontology Specifications. *Knowledge Acquisition, 5*(2), 199-220. http://dx.doi.org/10.1006/knac.1993.1008

Gruber, T. R. (1995). Toward principles for the design of ontologies used for knowledge sharing. *International Journal of Human-Computer Studies, 43*(5-6), 907-928. http://dx.doi.org/10.1006/ijhc.1995.1081

GTZ, WAJ, Dorsch Consult, & ConsulAqua Hamburg. (2009). Assessment of pump efficiency, pump operation and energy saving potential *Improvement of Energy Efficiency.* G.-J.-G. W. Program. Amman: GTZ/WAJ.

Guarino, N. (1995). Formal ontology, conceptual analysis and knowledge representation. *International Journal of Human-Computer Studies, 43*(5-6), 625-640. http://dx.doi.org/10.1006/ijhc.1995.1066

Gupta, H., Sorooshian, S., & Yapo, P. (1999). Status of Automatic Calibration for Hydrologic Models: Comparison with Multilevel Expert Calibration.

Journal of Hydrologic Engineering, 4(2), 135-143. http://dx.doi.org/10.1061/(ASCE)1084-0699(1999)4:2(135)

Gupta, J. N. D., & Harris, T. M. (1989). Decision support systems for small business. *J. SYST. MANAGE., 40*(2), 5

Gupta, V. K., & Sorooshian, S. (1983). Uniqueness and observability of conceptual rainfall-runoff model parameters: The percolation process examined. *Water Resour. Res., 19*(1), 269-276. http://dx.doi.org/10.1029/WR019i001p00269

Guttman, J. (2009). Hydrogeology. In H. Hötzl, P. Möller & E. Rosenthal (Eds.), *The Water of the Jordan Valley: Scarcity and Deterioration of Groundwater and Its Impact on the Regional Development*. Berlin, Heidelberg: Springer.

GWP. (2000). Integrated Water Resources Management (Background Paper No. 4) *TAC Background Papers*. Global Water Partnership-Technical Advisory Committee. Stockholm: Global Water Partnership.

Haag, D., & Kaupenjohann, M. (2001). Parameters, prediction, post-normal science and the precautionary principle—a roadmap for modelling for decision-making. *Ecological Modelling, 144*(1), 45-60. http://dx.doi.org/10.1016/s0304-3800(01)00361-1

Haddadin, M. J. (2002). *Diplomacy on the Jordan: international conflict and negotiated resolution*. Boston: Kluwer Academic Publishers.

Haddadin, M. J. (Ed.). (2006). *Water Resources in Jordan: Resources for the Future*. Washington: RFF Press.

Haddadin, P. M. J. (2006). *Water Resources in Jordan: Evolving Policies for Development, the Environment, and Conflict Resolution*: RFF Press.

Haettenschwiler, P. (1999). Neues anwenderfreundliches Konzept der Entscheidungsunterstützung. *Gutes Entscheiden in Wirtschaft, Politik und Gesellschaft*: vdf Hochschulverlag AG.

Hahne, K., Margane, A., & Borgstedt, A. (2008). Geological and Tectonic Setting of the Upper Wadi Shuayb Catchment Area, Jordan *Technical Cooperation Project: Groundwater Resources Management: Special Report*. B. MWI. Amman; Hannover.

Hajkowicz, S., & Collins, K. (2007). A Review of Multiple Criteria Analysis for Water Resource Planning and Management. *Water Resources Management, 21*(9), 1553-1566. http://dx.doi.org/10.1007/s11269-006-9112-5

Hamdi, M. R., Abu-Allaban, M., Al-Shayeb, A., Jaber, M., & Momani, M. N. (2009). Climate Change in Jordan: A Comprehensive Examination Approach. *Am. J. Environ. Sci., 5*, 10. http://dx.doi.org/10.3844/ajessp.2009.58.68

Hamner, J. H., & Wolf, A. T. (1998). Patterns in International Water Resource Treaties: The Transboundary Freshwater Dispute Database. *Colorado Journal of International Environmental Law and Policy, 9* (1997 Year Book), 20

Hareuveni, E., & Stein, Y. (2011). Dispossession and Exploitation: Israel's Policy in the Jordan Valley and Northern Dead Sea *B'Tselem Information Sheet*. B'Tselem. Jerusalem: B'Tselem.

Hargreaves, G. H., & Samani, Z. A. (1985). Reference Crop Evapotranspiration from Temperature. *Applied Engineering in Agriculture, 1*(2), 4

Helming, K., Pérez-Soba, M., & Tabbush, P. (2008). Sustainability Impact Assessment of Land Use Changes

Hicks, R., C. , Dattero, R., & Galup, S., D. (2006). The five-tier knowledge management hierarchy. *Journal of knowledge management, 10*(1), 19-31. http://dx.doi.org/10.1108/13673270610650076

Himpsl, K. (2007). *Wikis im Blended Learning : ein Werkstattbericht.* Boizenburg: vwh.

Hoff, H., Bonzi, C., Joyce, B., & Tielbörger, K. (2011). A Water Resources Planning Tool for the Jordan River Basin. *Water, 3*(3), 718-736

Hoffmann, R. (2008). A wiki for the life sciences where authorship matters. *Nature Genetics, 40*(9), 1047-1051. http://dx.doi.org/10.1038/ng.f.217

Holzner, B., & Marx, J. H. (1979). *Knowledge application: the knowledge system in society.* Boston: Allyn and Bacon.

Horowitz, A., & Flexer, K. (2001). *The Jordan Rift Valley.* Lisse: Balkema.

House, W. C. (1983). *Decision Support Systems : a data-based, model-oriented, user-developed discipline.* New York: Petrocelli.

HPC. (2009). The Demographic Opportunity in Jordan - A Policy Document. T. H. P. C. o. Jordan. Amman, Jordan: The Higher Population Council of Jordan.

Hubbard, K. G. (1994). Spatial Variability of Daily Weather Variables in the High-Plains of the USA. *Agricultural and Forest Meteorology, 68*(1-2), 29-41. http://dx.doi.org/10.1016/0168-1923(94)90067-1

Hughes, D. A., & Hannart, P. (2003). A desktop model used to provide an initial estimate of the ecological instream flow requirements of rivers in South Africa. *Journal of Hydrology, 270*(3–4), 167-181. http://dx.doi.org/10.1016/s0022-1694(02)00290-1

Humpal, D., El-Naser, H., Irani, K., Sitton, J., Renshaw, K., et al. (2012). A Review of Water Policies in Jordan and Recommendations for Strategic Priorities *Jordan Water Sector Assessment.* USAID. Amman: USAID.

IPCC. (2007). *Climate Change 2007: Synthesis Report. Contribution of Working Groups I, II and III to the Fourth Assessment Report of the Intergovernmental Panel on Climate Change.* Geneva: IPCC.

Itenfisu, D., Elliott, R. L., Allen, R. G., & Walter, I. A. (2003). Comparison of reference evapotranspiration calculations as part of the ASCE standardization effort. *Journal of Irrigation and Drainage Engineering-Asce, 129*(6), 440-448. http://dx.doi.org/10.1061/(Asce)0733-9437(2003)129:6(440)

IWAWaterWiki, (2012). IWA Water Wiki. Retrieved 2012-08-01, from http://www.iwawaterwiki.org

Izhikevich, E. (2006). Scholarpedia. *Scholarpedia, 1*, 1. http://dx.doi.org/10.4249/scholarpedia.1

Jacobs, K., Garfin, G., & Lenart, M. (2005). More than just talk: Connecting science and decision making. *Environment, 47*(9), 6-21. http://dx.doi.org/10.3200/ENVT.47.9.6-21

Jägerskog, A. (2003). Why states cooperate over shared water : the water negotiations in the Jordan River Basin. *PhD Thesis*, Linköpings universitet, Linköping.

Jamrah, A., Al-Futaisi, A., Prathapar, S., & Al Harrasi, A. (2008). Evaluating greywater reuse potential for sustainable water resources management in Oman. *Environmental Monitoring and Assessment, 137*(1-3), 315-327. http://dx.doi.org/10.1007/s10661-007-9767-2

Jasper, R., & Uschold, M. (1999). *A Framework for Understanding and Classifying Ontology Applications*. Paper presented at the IJCAI-99 Workshop on Ontologies and Problem-Solving Methods (KRR5).

Jeffrey, P., & Gearey, M. (2006). Integrated water resources management: lost on the road from ambition to realisation? *Water Science and Technology, 53*(1), 1-8. http://dx.doi.org/10.2166/wst.2006.001

JICA. (1995). The Study of Brackish Groundwater Desalination in Jordan. Japan International Cooperation Agency: Japan International Cooperation Agency.

Jiries, A., Ta'any, R., Abbassi, B., & Oroud, I. (2010). Agriculture Water Use Efficiency in Wadi Shu'eib Area, Jordan. *Pol. J. Environ. Stud., 19*(2), 5

Kaempgen, B., Riepl, D., Vrandecic, D., & Hermann, B. (in preparation). D203 - Definition of exchange format protocols for the connection of Dropedia with SMART-DB *SMART-Deliverables*. Karlsruhe: SMART Project.

Khlaifat, A., Hogan, M., Phillips, G., Nawayseh, K., Amira, J., et al. (2010). Long-term monitoring of the Dead Sea level and brine physico-chemical parameters "from 1987 to 2008". *Journal of Marine Systems, 81*(3), 207-212. http://dx.doi.org/10.1016/j.jmarsys.2009.11.005

Knorr-Cetina, K. D. (1982). Scientific Communities or Transepistemic Arenas of Research? A Critique of Quasi-Economic Models of Science. *Social Studies of Science, 12*(1), 101-130. http://dx.doi.org/10.1177/030631282012001005

Kok de, J.-L., & Wind, H. G. (2003). Design and application of decision-support systems for integrated water management: lessons to be learnt. *Physics and Chemistry of the Earth, Parts A/B/C, 28*(14-15), 571-578. http://dx.doi.org/10.1016/S1474-7065(03)00103-7

Kolkman, M. J., Kok, M., & van der Veen, A. (2005). Mental model mapping as a new tool to analyse the use of information in decision-making in integrated water management. *Physics and Chemistry of the Earth, 30*(4-5), 317-332. http://dx.doi.org/10.1016/j.pce.2005.01.002

Krabina, B. (2010). A Semantic Wiki on Cooperation in Public Administration in Europe. *Journal of Systemics, Cybernetics and Informatics, 8*(3)

Kristensen, P. (2004). *The DPSIR Framework*. Paper presented at the Workshop on a comprehensive assessment of the vulnerability of water resources to environmental change in Africa using river basin approach. .

Krötzsch, M., Vrandečić, D., & Völkel, M. (2006, November 5-9, 2006.). *Semantic MediaWiki*. Proceedings of the The Semantic Web - ISWC 2006, Athens, GA, USA.

Kuczera, G., & Mroczkowski, M. (1998). Assessment of hydrologic parameter uncertainty and the worth of multiresponse data. *Water Resour. Res., 34*(6), 1481-1489. http://dx.doi.org/10.1029/98wr00496

Kugler, S. (2010). *Non–Revenue-Water (NRW) Reduction – Key to improve economic performance of water and wastewater utilities*. Proceedings of the Arab Water Week, Amman.

Kühn, O., & Abecker, A. (1997). Corporate Memories for Knowledge Management in Industrial Practice: Prospects and Challenges. *Journal of Universal Computer Science, 3*(8), 25

Kunz, D. (2003). Soils in the Wadi Shueib catchment area and their protective potential for the groundwater - Salt area/Hashemite Kingdom of Jordan. *Unpublished Diploma-Thesis*, University of Karlsruhe, Karlsruhe.

Labadie, J. (2005). MODSIM: River basin managementdecision support system. In V. Singh & D. Frevert (Eds.), *Watershed Models*. Boca Raton, Florida: CRC Press.

Lafarge Cement Jordan. (2010). Annual Report and Financial Statements. L. C. Jordan. Amman: Lafarge Cement Jordan.

Lange, C. (2011). *Enabling collaboration on semiformal mathematical knowledge by semantic web integration*. Heidelberg and Amsterdam: AKA Verlag and IOS Press.

Lankford, B. A., Merrey, D. J., Cour, J., & Hepworth, N. (2007). *From Integrated to Expedient: An Adaptive Framework for River Basin Management in Developing Countries* (Vol. 110). Colombo, Sri Lanka: International Water Management Institute.

Lassila, O., & Swick, R. R. (1999). Resource Description Framework (RDF) Model and Syntax Specification Available from http://www.w3.org/TR/PR-rdf-syntax

Lein, Y., Hashhash, M. A., & H.A., H. (2001). Not Even A Drop: Water Crisis in Palestinian Villages. *B'Tselem Information Sheet*. B'Tselem. Jerusalem: B'Tselem.

Lenz, S. (1999). Hydrological Investigations along the Wadi al Kafrein and the Kafrein Reservoir ,Jordan. *Unpublished Diploma Thesis*, Universität Karlsruhe, Karlsruhe.

Leuf, B., & Cunningham, W. (2008). *The Wiki way : quick collaboration on the web* ([Nachdr.] ed.). Boston [u.a.]: Addison-Wesley.

Levite, H., Sally, H., & Cour, J. (2003). Testing water demand management scenarios in a water-stressed basin in South Africa: application of the WEAP model. *Physics and Chemistry of the Earth, 28*(20-27), 779-786. http://dx.doi.org/10.1016/j.pce.2003.08.025

Li, J. W., Robison, K., Martin, M., Sjodin, A., Usadel, B., et al. (2012). The SEQanswers wiki: a wiki database of tools for high-throughput sequencing analysis. *Nucleic Acids Research, 40*(D1), D1313-D1317. http://dx.doi.org/10.1093/Nar/Gkr1058

Liebowitz, J., & Wilcox, L. C. (1997). *Knowledge management and its integrative elements*. Boca Raton, Florida: CRC Press.

Lim, K. J., Engel, B. A., Tang, Z., Choi, J., Kim, K.-S., et al. (2005). Automated Web GIS Based Hydrograph Analysis Tool, WHAT. *JAWRA Journal of the American Water Resources Association, 41*(6), 1407-1416. http://dx.doi.org/10.1111/j.1752-1688.2005.tb03808.x

Lim, K. J., Park, Y. S., Kim, J., Shin, Y. C., Kim, N. W., et al. (2010). Development of genetic algorithm-based optimization module in WHAT system for hydrograph analysis and model application. *Computers & Geosciences, 36*(7), 936-944. http://dx.doi.org/10.1016/j.cageo.2010.01.004

Linstone, H. A., Turoff, M., & Helmer, O. (1975). *The Delphi method : techniques and applications*. Reading, Mass.; London: Addison-Wesley.

Liu, Y. Q., Gupta, H., Springer, E., & Wagener, T. (2008). Linking science with environmental decision making: Experiences from an integrated modeling approach to supporting sustainable water resources management. *Environmental Modelling & Software, 23*(7), 846-858. http://dx.doi.org/10.1016/j.envsoft.2007.10.007

Lo Presti, R., Barca, E., & Passarella, G. (2010). A methodology for treating missing data applied to daily rainfall data in the Candelaro River Basin (Italy). *Environmental Monitoring and Assessment, 160*(1-4), 1-22. http://dx.doi.org/10.1007/s10661-008-0653-3

LOD, Linked Open Data. (2012). Linked Data - Connect Distributed Data across the Web. Retrieved 04-09-2012, from http://linkeddata.org/

Loucks, D. P., Beek, E. v., & Stedinger, J. R. (2005). *Water resources systems planning and management : an introduction to methods, models and applications.* Paris: UNESCO.

Luijendijk, J., & Lincklaen Arriëns, W. T. (2007). *Bridging the knowledge gap: the value of knowledge networks.* Proceedings of the International Symposium Water for a Changing World - Developing Local Knowledge and Capacity, Delft, The Netherlands.

Lumsden, J., Hall, H., & Cruickshank, P. (2011). Ontology definition and construction, and epistemological adequacy for systems interoperability: A practitioner analysis. *Journal of Information Science, 37*(3), 246-253. http://dx.doi.org/10.1177/0165551511401804

Lykourentzou, I., Dagka, F., Papadaki, K., Lepouras, G., & Vassilakis, C. (2011). Wikis in enterprise settings: a survey. *Enterprise Information Systems, 6*(1), 1-53. http://dx.doi.org/10.1080/17517575.2011.580008

MacDonald, S. M., Hunting Technical Services Ltd., & Sultat al-Miyah al-Markaziyah. (1965). *East Bank Jordan water resources* (Vol. 6). London, Amman: Central Water Authority of Jordan.

Mahmoud, M., Liu, Y. Q., Hartmann, H., Stewart, S., Wagener, T., et al. (2009). A formal framework for scenario development in support of environmental decision-making. *Environmental Modelling & Software, 24*(7), 798-808. http://dx.doi.org/10.1016/j.envsoft.2008.11.010

Makridakis, S., Hogarth, R. M., & Gaba, A. (2009). Forecasting and uncertainty in the economic and business world. *International Journal of Forecasting, 25*(4), 18. http://dx.doi.org/10.1016/j.ijforecast.2009.05.012

Malczewski, J. (1999). *GIS and multicriteria decision analysis.* New York, NY: Wiley.

Marakas, G. M. (2003). *Decision support systems in the 21st century* (2. ed ed.). Upper Saddle River, NJ: Prentice Hall.

Margane, A., & BGR. (2002). *Contributions to the hydrogeology of Northern and Central Jordan.* Stuttgart: Schweizerbart.

Marie, A., & Vengosh, A. (2001). Sources of Salinity in Ground Water from Jericho Area, Jordan Valley. *Ground Water, 39*(2), 240-248. http://dx.doi.org/10.1111/j.1745-6584.2001.tb02305.x

Masri, M. (1963). The geology of the Amman-Zerqa area, Jordan. Amman: Central Water Authority.

Matthews, P. (1997). What Lies Beyond Knowledge Management: Wisdom Creation and Versatility. *Journal of Knowledge Management, 1*(3), 7. http://dx.doi.org/10.1108/EUM0000000004595

Matuszek, C., Cabral, J., Witbrock, M., & Deoliveira, J. (2006). An introduction to the syntax and content of Cyc. Retrieved from http://citeseerx.ist.psu.edu/viewdoc/summary?doi=10.1.1.126.4455

McCartney, M. P. (2000). The water budget of a headwater catchment containing a dambo. *Physics and Chemistry of the Earth, Part B: Hydrology, Oceans and Atmosphere, 25*(7–8), 611-616. http://dx.doi.org/10.1016/s1464-1909(00)00073-3

McIlwaine, S. (2009). Managing Jordan's Water Budget: Providing for Past, Present and Future Needs. In C. Lipchin, D. Sandler & E. Cushman (Eds.), *The Jordan River and Dead Sea Basin* (pp. 61-73): Springer Netherlands.

Medema, W. (2008). Integrated Water Resources Management And Adaptive Management: Shaping Science And Practice. *PhD Thesis*, Cranfield University, Cranfield.

Menzel, L., Teichert, E., & Weiss, M. (2007). *Climate change impact on the water resources of the semi-arid Jordan region.* Proceedings of the 3rd International Conference on Climate and Water, Helsinki.

Merrey, D. J. (2008). Is normative integrated water resources management implementable? Charting a practical course with lessons from Southern Africa. *Physics and Chemistry of the Earth, Parts A/B/C, 33*(8–13), 899-905. http://dx.doi.org/10.1016/j.pce.2008.06.026

Mietzner, D., & Reger, G. (2005). Advantages and disadvantages of scenario approaches for strategic foresight. *International Journal of Technology Intelligence and Planning, 1*(2), 220-239

Mikbel, S. H., & Zacher, W. (1981). The Wadi Shueib Structure in Jordan. *Neues Jahrb. Geol. Palaeontol., Monatshefte,* 5

Millington, P., Olson, D., & McMillan, S. (2006). An introduction to integrated river basin management *Integrated river basin management briefing note.* Washington D.C.: The World Bank. Retrieved from http://documents.worldbank.org/curated/en/2006/01/8513779/introduction-integrated-river-basin-management

Mitchell, B. (2004). Comments by Bruce Mitchell on Water Forum contribution: "Integrated Water Resources Management: A Reassessment" by Asit K. Biswas. *Water International, 29*(3), 398-399. http://dx.doi.org/10.1080/02508060408691794

Mittra, S. S. (1986). *Decision support systems: tools and techniques.* New York: Wiley.

Molle, F. (2008). Nirvana Concepts, Narratives and Policy Models: Insights from the Water Sector. *Water Alternatives, 1*(1), 25

Monteith, J. L. (1973). *Principles of environmental physics.* London: Arnold.

Moreda, F. (1999). Conceptual rainfall - runoff models for different time steps with special consideration for semi-arid and arid catchments. *PhD-Thesis*, Vrieje Universiteit Brussel, Brussel.

Moriasi, D. N., Arnold, J. G., Van Liew, M. W., Bingner, R. L., Harmel, R. D., et al. (2007). Model evaluation guidelines for systematic quantification of accuracy in watershed simulations. *Transactions of the Asabe, 50*(3), 16

Motik, B., Patel-Schneider, P. F., Parsia, B., Bock, C., Fokoue, A., et al. (2009). OWL 2 Web Ontology Language: Structural Specification and Functional-

Style Syntax *W3C Recommendation*. B. Boris Motik, P. F. Patel-Schneider & B. Parsia: W3C. Retrieved from http://www.w3.org/TR/owl2-syntax/

Moxey, A., Whitby, M., & Lowe, P. (1998). Agri-environmental indicators: issues and choices. *Land Use Policy, 15*(4), 265-269. http://dx.doi.org/10.1016/S0264-8377(98)00023-4

Mukhtarov, F. G. (2008). Intellectual history and current status of Integrated Water Resources Management: A global perspective. In C. Pahl-Wostl, P. Kabat & J. Möltgen (Eds.), *Adaptive and integrated water management: coping with complexity and uncertainty* (pp. 167-185). Berlin Heidelberg: Springer

Müller, F., & Wiggering, H. (2004). Erfahrungen und Entwicklungspotentiale von Ziel- und Indikatorensystemen. In H. Wiggering, F. Müller, M. Huch & H. Geldmacher (Eds.), *Umweltziele und Indikatoren* (pp. 221-234): Springer Berlin Heidelberg.

Müller, R. A., Cardona, J., Subah, A., Abbassi, B., & van Afferden, M. (2011, August 21-27). *Implementation research for decentralized wastewater concepts in Jordan*. Proceedings of the World Water Week, Stockholm.

Murakami, M. (1998). Alternative strategies in the inter-state regional development of the Jordan Rift Valley. In I. Kobori & M. H. Glantz (Eds.), *Central Eursian Water Crisis: Caspian, Aral, and Dead Seas*: United Nations University Press.

MW, MediaWiki. (2012). User Manual: Editing. Retrieved 2012-05-22, from http://www.mediawiki.org/wiki/Help:Editing

MWI. (1997). *Water Strategy of Jordan*. Amman:Ministry for Water and Irrigation.

MWI. (2004). National Water Master Plan for Jordan. MWI. Amman: Ministry for Water and Irrigation, .

MWI. (2009). The Hashemite Kingdom of Jordan: Jordan's second national communication to the United Nations Framework Convention on Climate Change. *National Communications to the United Nations Framework Convention on Climate Change*. M. o. W. a. Irrigation. Amman: MWI.

MWI. (2010). Annual Report. Ministry of Water and Irrigation. Amman.

Mysiak, J., Giupponi, C., & Rosato, P. (2005). Towards the development of a decision support system for water resource management. *Environmental Modelling & Software, 20*(2), 203-214. http://dx.doi.org/10.1016/j.envsoft.2003.12.019

Nakicenovic, N., Alcamo, J., Davis, G., de Vries, B., Fenhann, J., et al. (2000). *Special Report on Emissions Scenarios : a special report of Working Group III of the Intergovernmental Panel on Climate Change*. New York, NY: Cambridge University Press.

Nash, J. E., & Sutcliffe, J. V. (1970). River flow forecasting through conceptual models part I — A discussion of principles. *Journal of Hydrology, 10*(3), 282-290. http://dx.doi.org/10.1016/0022-1694(70)90255-6

Nazer, D. W., Siebel, M. A., Van der Zaag, P., Mimi, Z., & Gijzen, H. J. (2008). Water Footprint of the Palestinians in the West Bank1. *JAWRA Journal of the American Water Resources Association, 44*(2), 449-458. http://dx.doi.org/10.1111/j.1752-1688.2008.00174.x

Newman, B. D., & Conrad, K. W. (2000, 30-31 Oct. 2000). *A Framework for Characterizing Knowledge Management Methods, Practices, and Technologies*. Proceedings of the Third Int. Conf. on Practical Aspects of Knowledge Management (PAKM2000), Basel, Switzerland.

Nguyen, T. G., de Kok, J. L., & Titus, M. J. (2007). A new approach to testing an integrated water systems model using qualitative scenarios. *Environmental Modelling & Software, 22*(11), 1557-1571. http://dx.doi.org/10.1016/j.envost.2006.08.005

Nielsen, J., (2006). Participation Inequality: Encouraging More Users to Contribute. *Jakob Nielsen's Alertbox*, from http://www.useit.com/alertbox/participation_inequality.html

Niemeijer, D., & de Groot, R. S. (2008). A conceptual framework for selecting environmental indicator sets. *Ecological Indicators, 8*(1), 14-25. http://dx.doi.org/10.1016/j.ecolind.2006.11.012

Nonaka, I., & Takeuchi, H. (1995). *The knowledge creating company : how Japanese companies create the dynamics of innovation*. New York, NY: Oxford Univ. Press.

Nusier, O., Alawneh, A., & Malkawi, A. (2002). Remedial measures to control seepage problems in the Kafrein dam, Jordan. *Bulletin of Engineering Geology and the Environment, 61*(2), 145-152. http://dx.doi.org/10.1007/s100640100131

O'Dell, C. S., Grayson, C. J., & Essaides, N. (1998). *If only we knew what we know: the transfer of internal knowledge and best practice*. New York, NY: Free Press.

O'Leary, D. E. (2008). Wikis: 'From Each According to His Knowledge'. *IEEE Computer, 41*(2), 34-41. http://dx.doi.org/10.1109/mc.2008.68

Oberle, D., Ankolekar, A., Hitzler, P., Cimiano, P., Sintek, M., et al. (2007). DOLCE ergo SUMO: On foundational and domain models in the SmartWeb integrated ontology (SWIntO). *Journal of Web Semantics, 5*(3), 156-174. http://dx.doi.org/10.1016/j.websem.2007.06.002

OECD. (1993). OECD core set of indicators for environmental performance reviews *Environment Monographs*. OECD. Paris: Organistation for Economic Co-Operation and Development.

OECD. (2006). Applying Strategic Environmental Assessment. Good Practice Guidance for Development Co-Operation *DAC Guidelines and Reference Series*. OECD: Organisation for Economic Co-Operation and Development.

Oki, T., & Kanae, S. (2006). Global hydrological cycles and world water resources. *Science, 313*(5790), 1068-1072. http://dx.doi.org/10.1126/science.1128845

Oki, T., Valeo, C., & Heal, K. (2006). *Hydrology 2020 : an integrating science to meet world water challenges*. Wallingford: IAHS Press.

Oren, E., Völkel, M., Breslin, J., & Decker, S. (2006). Semantic Wikis for Personal Knowledge Management Database and Expert Systems Applications. In S. Bressan, J. Küng & R. Wagner (Eds.), (Vol. 4080, pp. 509-518): Springer Berlin / Heidelberg.

Pahl-Wostl, C. (2007). Transitions towards adaptive management of water facing climate and global change. *Water Resources Management, 21*(1), 49-62. http://dx.doi.org/10.1007/s11269-006-9040-4

Pahl-Wostl, C., Downing, T., Kabat, P., Magnuszewski, P., Meigh, J., et al. (2005). Transition to adaptive water management: The NeWater project *NeWater Working Paper: New approaches to adaptive water management under uncertainty*. I. o. E. S. Research. Osnabruck: University of Osnabruck.

Palma, R., Corcho, O., Gómez-Pérez, A., & Haase, P. (2011). A holistic approach to collaborative ontology development based on change management. *Web Semantics: Science, Services and Agents on the World Wide Web, 9*(3), 299-314. http://dx.doi.org/10.1016/j.websem.2011.06.007

Park, T. K. (2011). The visibility of Wikipedia in scholarly publications. *First Monday, 16*(8). Retrieved from http://www.firstmonday.org/htbin/cgiwrap/bin/ojs/index.php/fm/article/view/3492/3031

Parker, D. H. (1970). *Investigations of the Sandstone Aquifers of East Jordan: The Hydrogeology of the Mesozoic-cainozoic Aquifers of the Western Highlands and Plateau of East Jordan*: F.A.O. Documentation Centre.

Parker, K. R., & Chao, J. T. (2007). Wiki as a Teaching Tool. *Interdisciplinary Journal of Knowledge and Learning Objects, 3*, 15

Penman, H. L. (1948). Natural Evaporation from Open Water, Bare Soil and Grass. *Proceedings of the Royal Society of London. Series A. Mathematical and Physical Sciences, 193*(1032), 120-145. http://dx.doi.org/10.1098/rspa.1948.0037

Phillips, D. J. H., Attili, S., McCaffrey, S., & Murray, J. S. (2004). Factors Relating to the Equitable Distribution of Water in Israel and Palestine. In H. I. Shuval & H. Dwiek (Eds.), *Water resources in the Middle East: the Israeli-Palestinian water issues : from conflict to cooperation*: Springer.

Polanyi, M. (1966). The Logic of Tacit Inference. *Philosophy, 41*(155), 1-18. http://www.jstor.org/stable/3749034

Power, D. J. (2002). *Decision support systems : concepts and resources for managers*. Westport, CT: Quorum Books.

Priestley, C., & Taylor, R. J. (1972). Assessment of Surface Heat-Flux and Evaporation Using Large-Scale Parameters. *Monthly Weather Review, 100*(2), 3. http://dx.doi.org/10.1175/1520-0493(1972)100<0081:OTAOSH>2.3.CO;2

Quennell, A. M. (1951). *The geology and mineral resources of (former) Trans-Jordan* (Vol. 2): Colonial Geology and Mineral Resources.

Rahaman, M. M., & Varis, O. (2005). Integrated water resources management: evolution, prospects and future challenges. *Sustainability: Science, Practice, & Policy, 1*(1), 6

Rapport, D. J., & Friend, A. M. (1979). Towards a Comprehensive Framework for Environmental Statistics: A Stress-Response Approach. *Statistics Canada, Occasional Papers*, 90

RCW, & MWI. (2009). *Water for Life: Jordan's Water Strategy 2008-2022*. Amman:Ministry for Water and Irrigation.

RDFS, RDF Schema. (2010). RDF Vocabulary Description Language 1.0: RDF Schema. *W3C Recommendation*, from http://www.w3.org/TR/rdf-schema/

Reed, S. L., & Lenat, D. B. (2002). Mapping Ontologies into Cyc *AAAI Technical Report WS-02-11*: AAAI (www.aaai.org).

Refsgaard, J. C. (1997). Parameterisation, calibration and validation of distributed hydrological models. *Journal of Hydrology, 198*(1–4), 69-97. http://dx.doi.org/10.1016/s0022-1694(96)03329-x

Refsgaard, J. C., Henriksen, H. J., Harrar, W. G., Scholten, H., & Kassahun, A. (2005). Quality assurance in model based water management – review of existing practice and outline of new approaches. *Environmental Modelling & Software, 20*(10), 1201-1215. http://dx.doi.org/10.1016/j.envsoft.2004.07.006

Riepl, D., Kaempgen. B, & Wolf, L. (2011). *Making informed decisions - a collaborative and knowledge based IWRM planning exercise in Wadi Shueib, Jordan.* Proceedings of the IWRM2011, Dresden, Germany.

Rijsberman, F. R. (2006). Water scarcity: Fact or fiction? *Agricultural Water Management, 80*(1–3), 5-22. http://dx.doi.org/10.1016/j.agwat.2005.07.001

Ritchey, T. (2006). Problem structuring using computer-aided morphological analysis. *Journal of the Operational Research Society, 57*(7), 792-801. http://dx.doi.org/10.1057/palgrave.jors.2602177

Rittel, H. W. J., & Webber, M. M. (1973). Dilemmas in a general theory of planning. *Policy Sciences, 4*(2), 155-169. http://dx.doi.org/10.1007/bf01405730

Roelof, P. u. B. (1999). Questions in knowledge management: defining and conceptualising a phenomenon. *Journal of knowledge management, 3*(2), 94-110. http://dx.doi.org/10.1108/13673279910275512

Rogers, P., Bhatia, R., & Huber-Lee, A. (1998). Water as a Social and Economic Good: How to Put the Principle into Practice *TAC Background Papers*. G. W. Partnership. Stockholm: Global Water Partnership.

Rosenberg, D., & Lund, J. (2009). Modeling Integrated Decisions for a Municipal Water System with Recourse and Uncertainties: Amman, Jordan. *Water Resources Management, 23*(1), 85-115. http://dx.doi.org/10.1007/s11269-008-9266-4

Rosenthal, R. (1979). The File Drawer Problem and Tolerance for Null Results. *Psychological Bulletin, 86*(3), 638-641. http://dx.doi.org/10.1037//0033-2909.86.3.638

Rothenberger, D. (2009). Improving Water Utility Performance Through Local Private Sector Participation - *Discussion Paper Series of the German-Jordanian Programme: Management of Water Resources*. Amman: GTZ.

Roux, D. J., Rogers, K. H., Biggs, H. C., Ashton, P. J., & Sergeant, A. (2006). Bridging the science-management divide: Moving from unidirectional knowledge transfer to knowledge interfacing and sharing. *Ecology and Society, 11*(1). http://hdl.handle.net/10204/953

Rowley, J. (2007). The wisdom hierarchy: representations of the DIKW hierarchy. *Journal of Information Science, 33*(2), 163-180. http://dx.doi.org/10.1177/0165551506070706

Roy, B. (1968). Classement et choix en présence de points de vue multiples (la méthode ELECTRE). *La Revue d'Informatique et de Recherche Opérationelle (RIRO), 8*, 18

Roy, B. (1996). *Multicriteria methodology for decision aiding.* Dordrecht ; London: Kluwer.

Saaty, R. W. (1987). The analytic hierarchy process—what it is and how it is used. *Mathematical Modelling, 9*(3–5), 161-176. http://dx.doi.org/10.1016/0270-0255(87)90473-8

Sahawneh, J. (2011). Structural Control of Hydrology, Hydrogeology and Hydrochemistry along the Eastern Escarpment of the Jordan Rift Valley, JORDAN. *PhD-Thesis*, Karlsruhe Institut für Technologie, Karlsruhe.

Saidam, M. Y., & Ibrahim, M. N. (2006). Institution and Policy Framework - Analysis of the Water Sector in Jordan *Co-ordination Action for Autonomous Desalination Units based on Renewable Energy Systems*. R. S. Society. Amman: Royal Scientific Society.

Salameh, E. (1980). The Suweileh Structure. *N. Jb. Geol. Palaeol., Monatshefte*(7), 10

Salameh, E. (2008). State of Water Strategy and Policy. In H. Hötzl, P. Möller & E. Rosenthal (Eds.), *The Water of the Jordan Valley*. Berlin Heidelberg: Springer-Verlag.

Salameh, E., & El-Naser, H. (1999). Does the Actual Drop in Dead Sea Level Reflect the Development of Water Sources Within its Drainage Basin? *Acta hydrochimica et hydrobiologica, 27*(1), 5-11. http://dx.doi.org/10.1002/(sici)1521-401x(199901)27:1<5::aid-aheh5>3.0.co;2-z

Salameh, E., & Udluft, P. (1985). *The hydrodynamic pattern of the central part of Jordan* (Vol. c38). Hannover.

Saqr, R. (2001). Water Treatment Plant reopens after $7.5m US-funded rehabilitation. *Jordan Times*, Wednesday, November 14, 2001.

Schacter, D. L., Addis, D. R., & Buckner, R. L. (2007). Remembering the past to imagine the future: the prospective brain. *Nat Rev Neurosci, 8*(9), 657-661. http://dx.doi.org/10.1038/nrn2213

Schaffert, S., Eder, J., Grünwald, S., Kurz, T., & Radulescu, M. (2009). KiWi – A Platform for Semantic Social Software (Demonstration). In L. Aroyo, P. Traverso, F. Ciravegna, P. Cimiano, T. Heath, E. Hyvönen, R. Mizoguchi, E. Oren, M. Sabou & E. Simperl (Eds.), *The Semantic Web: Research and Applications* (Vol. 5554, pp. 888-892): Springer Berlin / Heidelberg.

Schaffert, S., Gruber, A., & Westenthaler, R. (2005, November, 23-25, 2005). *A Semantic WIKI for Collaborative Knowledge Formation*. Proceedings of the SEMANTICS 2005, Vienna, Austria.

Scheffer, J. (2002). Dealing with Missing Data. *Res. Lett. Inf. Math. Sci. , 3*, 7

Schoemaker, P. J. H. (1995). Scenario Planning: A Tool for Strategic Thinking. *Sloan Management Review, 36*(2), 15

Scholten, H. M. (2008). Better Modelling Practice: An Ontological Perspective on Multidisciplinary, Model-based Problem Solving. *PhD-Thesis*, Wageningen Universiteit. Available from http://books.google.de/books?id=5BZTPgAACAAJ

Schumm, S. A. (1977). *Drainage basin morphology*. Stroudsburg, Penns.: Dowden.

Schwartz, P. (1991). *The art of the long view* (1st ed ed.). New York, NY: Currency Doubleday.

Scott, D. A. (Ed.). (1995). *A directory of wetlands in the Middle East*. Gland, Switzerland: IUCN, The World Conservation Union ; Slimbridge : International Waterfowl and Wetlands Research Bureau.

Segnestam, L. (2002). Indicators of Environment and Sustainable Development - Theories and Practical Experience *Environmental Economics Series*. Washington, D.C.: The World Bank.

Segura, F. S., Pardo-Pascual, J. E., Rossello, V. M., Fornos, J. J., & Gelabert, B. (2007). Morphometric indices as indicators of tectonic, fluvial and karst processes in calcareous drainage basins, South Menorca Island, Spain. *Earth Surface Processes and Landforms, 32*(13), 1928-1946. http://dx.doi.org/10.1002/Esp.1506

Segura/IP3 Partners LLC. (2009). Pricing of Water and Wastewater Services in Amman and Subsidy Options: Conceptual Framework, Recommendations, and Pricing Model. *Final report*. USAID. Amman: USAID.

Servat, E., & Dezetter, A. (1993). Rainfall-runoff modelling and water resources assessment in northwestern Ivory Coast. Tentative extension to ungauged catchments. *Journal of Hydrology, 148*(1–4), 231-248. http://dx.doi.org/10.1016/0022-1694(93)90262-8

Shadbolt, N., Hall, W., & Berners-Lee, T. (2006). The Semantic Web revisited. *Ieee Intelligent Systems, 21*(3), 96-101. http://dx.doi.org/10.1109/Mis.2006.62

Shepard, D. (1968). *A two-dimensional interpolation function for irregularly-spaced data*. Paper presented at the Proceedings of the 1968 23rd ACM national conference.

Simon, H. A. (1960). *The new science of management decision*. New York: Harper & Row.

Simperl, E. (2009). Reusing ontologies on the Semantic Web: A feasibility study. *Data & Knowledge Engineering, 68*(10), 20. http://dx.doi.org/10.1016/j.datak.2009.02.002

Singhal, B. B. S., & Gupta, R. P. (2010). Applied Hydrogeology of Fractured Rocks Available from http://dx.doi.org/10.1007/978-90-481-8799-7

Smakhtin, V., Revenga, C., & Döll, P. (2004). A Pilot Global Assessment of Environmental Water Requirements and Scarcity. *Water International, 29*(3), 307-317. http://dx.doi.org/10.1080/02508060408691785

Smeets, E., & Weterings, R. (1999). Environmental indicators : typology and overview *Technical Report*. Copenhagen: European Environment Agency.

Smith, B., Kusnierczyk, W., Schober, D., & Ceusters, W. (2006, 2006). *Towards a Reference Terminology for Ontology Research and Development in the Biomedical Domain*. Proceedings of the KR-MED 2006 "Biomedical Ontology in Action", Baltimore, Maryland, USA.

Smith, C. G. (1966). The Disputed Waters of the Jordan. *Transactions of the Institute of British Geographers*(40), 111-128

SMW, Semantic MediaWiki. (2012). User Manual. Retrieved 2012-05-22, from http://semantic-mediawiki.org/wiki/Help:User_manual

Solis, C., Ali, N., & Babar, M. A. (2009, 19-19 May 2009). *A Spatial Hypertext Wiki for Architectural Knowledge Management*. Proceedings of the Wikis for Software Engineering, 2009. WIKIS4SE '09.

Sommaripa, L. (2011). Water Public Expenditure Perspective Working Paper *Jordan Fiscal Reform II Project*. USAID. Amman: USAID.

Sprague, R. H., & Carlson, E. D. (1982). *Building effective decision support systems*. Englewood Cliffs, N.J.: Prentice-Hall.

Sprague, R. H., & Watson, H. J. (1986). *Decision Support Systems : Putting theory into practice.* Englewood Cliffs, N. J.: Prentice-Hall.

SQUID, (2012). squid-cache.org. Retrieved 2012-02-12, from http://www.squid-cache.org/

Srdjevic, B. (1987). Identification of the Control Strategies in Water Resources Systems with Reservoirs by Use of Network Models. *Ph. Dissertation*, University of Novi Sad, Yugoslavia.

Storz, R. (2004). GIS-based groundwater hazard mapping of the Wadi Shueib catchment area, Jordan. *Umpublished Diploma-Thesis*, University of Karlsruhe, Karlsruhe.

Studer, R., Benjamins, V. R., & Fensel, D. (1998). Knowledge Engineering: Principles and methods. *Data & Knowledge Engineering, 25*(1-2), 161-197. http://dx.doi.org/10.1016/S0169-023x(97)00056-6

Suleiman, R. (2004). The Historical Evolution of the Water Resources. Development in the Jordan River Basin in Jordan *MREA-IWMI Working Papers*. MREA-IWMI. Amman: MREA-IWMI.

SWIVT, Semantic Wiki Vocabulary and Terminology. (2010). SWIVT Ontology Specification. Retrieved 2012-08-01, from http://semantic-mediawiki.org/swivt/

Ta'any, R. A. A. (1992). Hydrological and Hydrochemical Study of the Major Springs in the Wadi Shueib Catchment Area. *Master-Thesis*, Yarmouk University, Irbid.

Tabios, G. Q., & Salas, J. D. (1985). A Comparative Analysis of Techniques for Spatial Interpolation of Precipitation. *JAWRA Journal of the American Water Resources Association, 21*(3), 365-380. http://dx.doi.org/10.1111/j.1752-1688.1985.tb00147.x

Tallaksen, L. M. (1995). A Review of Baseflow Recession Analysis. *Journal of Hydrology, 165*(1-4), 349-370. http://dx.doi.org/10.1016/0022-1694(94)02540-R

Tchobanoglous, G., Burton, F. L., Stensel, H. D., Metcalf, & Eddy. (2003). *Wastewater Engineering: Treatment and Reuse*: McGraw-Hill.

Thakker, D., Dimitrova, V., Lau, L., Denaux, R., Karanasios, S., et al. (2011). *A priori ontology modularisation in ill-defined domains*. Paper presented at the 7th International Conference on Semantic Systems.

Thornthwaite, C. W. (1948). An Approach toward a Rational Classification of Climate. *Geographical Review, 38*(1), 55-94

Timpe, C., & Scheepers, M. J. J. (2003). A Look into the Future: Scenarios for Distributed Generation in Europe *Policy and Regulatory Roadmaps for the Integration of Distributed Generation and the Development of Sustainable Electricity Networks*: SUSTELNET.

Trappe, J. (2007). Delineation of Groundwater Protection Zones for Baqqouria Spring According to Jordanian Guidelines. *MSc thesis*, FH Koeln, Cologne.

Tulving, E. (1985). Memory and consciousness. *Canadian Psychology, 26*(1), 12. http://dx.doi.org/10.1037/h0080017

Tuomi, I. (1999). Data is more than knowledge: Implications of the reversed knowledge hierarchy for knowledge management and organizational memory. *Journal of Management Information Systems, 16*(3), 103-117

Turc, L. (1961). Estimation of irrigation water requirements, potential evapotranspiration: a simple climatic formula evolved up to date. *Annals of Agronomy, 12*, 36. http://dx.doi.org/10.1109/HICSS.1999.772795

Twort, A. C., Ratnayaka, D. D., & Brandt, M. J. (2000). *Water supply* (5. ed ed.). London: Arnold [u.a.].

UN-Water, & FAO. (2007). Coping with Water Scarcity. Challenge of the twenty-first century. *World Water Day 2007 Reports*: Food and Agriculture Organization of the United Nations (FAO) & UN-Water Retrieved from http://www.fao.org/nr/water/docs/escarcity.pdf (accessed 10 December 2011)

UN. (2001). Indicators of Sustainable Development: Guidelines and Methodologies *Background Paper* Dept. of Economic & Social Affairs of the United Nations. Retrieved from http://books.google.de/books?id=r2gRCCNR_egC

UN. (2010a). The Millennium Development Goals Report 2010 *The Millenium Development Goal Reports*. New York, NY: United Nations Department of Economic and Social Affairs.

UN. (2010b). World Population Prospects: The 2010 Revision. Population Division: Department of Economic and Social Affairs of the United Nations. Retrieved from http://esa.un.org/unpd/wpp/unpp/panel_population.htm

UNEP/GRID-Arendal. (2008). Areas of physical and economic water scarcity.: UNEP/GRID-Arendal Maps and Graphics Library.

United Nations. Development Programme. (2006). *Beyond scarcity : power, poverty and the global water crisis*. New York, NY: UNDP.

US-SCS. (1972). *US National Soil Conservation Service: National Engineering Handbook Section 4 Hydrology*. [S.l.]: U.S.G.P.O.

USAID. (2002). Integratet Water Resources Management - Ushering in a Blue Revolution: USAID.

USAID/JISSP. (2011). Conceptual Update to Water for Life, Jordan's Water Strategy 2008-2022. Amman: Ministry for Water and Irrigation.

van Afferden, M., Cardona, J. A., Rahman, K. Z., Daoud, R., Headley, T., et al. (2010). A step towards decentralized wastewater management in the Lower Jordan Rift Valley. *Water Science and Technology, 61*(12), 3117-3128. http://dx.doi.org/10.2166/wst.2010.234

van der Heijden, K. (1996). *Scenarios: the art of strategic conversation*. Chichester, England New York: John Wiley & Sons.

van der Merwe, L. (2008). Scenario-Based Strategy in Practice: A Framework. *Advances in Developing Human Resources, 10*(2), 216-239. http://dx.doi.org/10.1177/1523422307313321

van der Molen, I., Hasan, W., & Tamimi, A. (2011). Multi-stakeholder processes, service delivery and state institutions:Experiences from the West Bank and Gaza Strip, Palestinian Territories: Peace, Security and Development Network.

van der Zaag, P. (2005). Integrated Water Resources Management: Relevant concept or irrelevant buzzword? A capacity building and research agenda for Southern Africa. *Physics and Chemistry of the Earth, Parts A/B/C, 30*(11–16), 867-871. http://dx.doi.org/10.1016/j.pce.2005.08.032

Vandewiele, G. L., & Ni-Lar-Win. (1998). Monthly water balance models for 55 basins in 10 countries. *Hydrological Sciences Journal-Journal Des Sciences Hydrologiques, 43*(5), 687-699. http://dx.doi.org/10.1023/A:1007916816469

Vogel, R. M., & Sankarasubramanian, A. (2003). Validation of a watershed model without calibration. *Water Resources Research, 39*(10). http://dx.doi.org/10.1029/2002wr001940

Volk, M., Hirschfeld, J., Schmidt, G., Bohn, C., Dehnhardt, A., et al. (2007). A SDSS-based Ecological-economic Modelling Approach for Integrated River Basin Management on Different Scale Levels – The Project FLUMAGIS. *Water Resources Management, 21*(12), 2049-2061. http://dx.doi.org/10.1007/s11269-007-9158-z

Völkel, M., Krötzsch, M., Vrandecic, D., Haller, H., & Studer, R. (2006). *Semantic Wikipedia*. Paper presented at the 15th international conference on World Wide Web.

Vrandecic, D., Pinto, S., Tempich, C., & Sure, Y. (2005). The DILIGENT knowledge processes. *Journal of knowledge management, 9*(5), 85-96. http://dx.doi.org/10.1108/13673270510622474

Wack, P. (1985). *Scenarios: Uncharted Waters Ahead*: Harvard Business Review.

Wagener, T., & Wheater, H. S. (2006). Parameter estimation and regionalization for continuous rainfall-runoff models including uncertainty. *Journal of Hydrology, 320*(1–2), 132-154. http://dx.doi.org/10.1016/j.jhydrol.2005.07.015

Wagner, C. (2006). Breaking the knowledge acquisition bottleneck through conversational knowledge management. *Innovative Technologies for Information Resources Management, 19*(1), 200. http://dx.doi.org/10.4018/irmj.2006010104

Waldrop, M. M. (2008). Information technology - Science 2.0. *Scientific American, 298*(5), 68-73. http://dx.doi.org/10.1038/scientificamerican0508-68

Wallace, D. P. (2007). *Knowledge management : historical and cross-disciplinary themes*. Westport, Conn.: Libraries Unlimited.

Wang, C. L., & Ahmed, P. K. (2005). The knowledge value chain: a pragmatic knowledge implementation network. *Handbook of Business Strategy, 6*(1), 321-326. http://dx.doi.org/10.1108/08944310510558115

WaterWiki, (2012). WaterWiki. Retrieved 2012-01-08, from http://waterwiki.net

Weinberger, G., Livshitz, Y., Givati, A., Zilberbrand, M., Tal, A., et al. (2012). The Natural Water Resources Between the Mediterranean Sea and the Jordan River. Jerusalem: Israel Hydrological Service.

Werz, H. (2006). *The use of remote sensing imagery for groundwater risk intensity mapping in the Wadi Shueib, Jordan* Karlsruhe: PhD Thesis at the Fakultät für Bauingenieur-, Geo- und Umweltwissenschaften - Universität Karlsruhe.

Wikia, (2012). Wikia Free Wiki Hosting. Retrieved 2012-08-01, from http://www.wikia.com

Wikidata, (2012). Wikidata - The free knowledge base that anyone can edit. Retrieved 2012-01-11, from http://www.wikidata.org/wiki/Wikidata:Main_Page

Wikimatrix, (2012). List of Wiki-Engines. Retrieved 2011-10-10, from http://www.wikimatrix.org

Williams, A. J. (2008). Internet-based tools for communication and collaboration in chemistry. *Drug Discovery Today, 13*(11-12), 502-506. http://dx.doi.org/10.1016/j.drudis.2008.03.015

Wolf, A. T., Natharius, J. A., Danielson, J. J., Ward, B. S., & Pender, J. K. (1999). International River Basins of the World. *International Journal of Water Resources Development, 15*(4), 387-427. http://dx.doi.org/10.1080/079900629948682

Wolf, L., & Hötzl, H. (2011). *SMART-IWRM: Integrated water resources management in the lower Jordan Rift Valley : project report phase I* (Print on demand ed.). Karlsruhe: KIT Scientific Publishing.

Wolf, L., Riepl, D., Subah, A., Tamimi, A., Guttman, J., et al. (submitted). IWRM and Implementation - Reviewing results and uptake from a large-scale trans-boundary research project at the Lower Jordan River. *Water, Science & Technology*

Woods, E., & Sheina, M. (1999). *Knowledge management: building the collaborative enterprise.* London: Ovum.

World Bank. (2009a). Red Sea - Dead Sea Water Conveyance Study Program Feasibility Study: Options Screening and Evaluation Report: Executive Summary *Red Sea - Dead Sea Water Conveyance Study Program Feasibility Study*: The World Bank.

World Bank. (2009b). West Bank and Gaza - Assessment of Restrictions on Palestinian Water Sector Development *Sector Note*. Washington: The World Bank. Retrieved from http://go.worldbank.org/XFU3CZY1T0

WP, Wikipedia. (2012a). Open Data. Retrieved 2012-10-01, from http://en.wikipedia.org/wiki/Open_data

WP, Wikipedia. (2012b). Open Government. Retrieved 2012-01-10, from http://en.wikipedia.org/wiki/Open_government

WP, Wikipedia. (2012c). Open Notebook Science. Retrieved 2012-10-01, from http://en.wikipedia.org/wiki/Open_notebook_science

WP, Wikipedia. (2012d). Science 2.0. Retrieved 2012-01-10, from http://en.wikipedia.org/wiki/Science_2.0

Xu, C. Y., & Singh, V. P. (1998). A Review on Monthly Water Balance Models for Water Resources Investigations. *Water Resources Management, 12*(1), 20-50. http://dx.doi.org/10.1023/a:1007916816469

Yates, D., & Paquette, S. (2011). Emergency knowledge management and social media technologies: A case study of the 2010 Haitian earthquake. *International Journal of Information Management, 31*(1), 6-13. http://dx.doi.org/10.1016/j.ijinfomgt.2010.10.001

Yates, D., Purkey, D., Sieber, J., Huber-Lee, A., Galbraith, H., et al. (2009). Climate Driven Water Resources Model of the Sacramento Basin, California. *Journal of Water Resources Planning and Management-Asce, 135*(5), 303-313. http://dx.doi.org/10.1061/(Asce)0733-9496(2009)135:5(303)

Yates, D., Sieber, J., Purkey, D., & Huber-Lee, A. (2005). WEAP21 - A demand-, priority-, and preference-driven water planning model Part 1: Model characteristics. *Water International, 30*(4), 487-500. http://dx.doi.org/10.1080/02508060508691893

Yates, D. N., & Strzepek, K. M. (1998). Modeling the Nile Basin under Climatic Change. *Journal of Hydrologic Engineering, 3*(2), 98-108. http://dx.doi.org/10.1061/(ASCE)1084-0699(1998)3:2(98)

Ye, W., Bates, B. C., Viney, N. R., Sivapalan, M., & Jakeman, A. J. (1997). Performance of conceptual rainfall-runoff models in low-yielding ephemeral catchments. *Water Resources Research, 33*(1), 153-166. http://dx.doi.org/10.1029/96wr02840

Yoffe, S., Wolf, A. T., & Giordano, M. (2003). Conflict and Cooperation over international freshwater respurces: Indicators of Basins at risk. *JAWRA Journal of the American Water Resources Association, 39*(5), 1109-1126. http://dx.doi.org/10.1111/j.1752-1688.2003.tb03696.x

Young, J., Rinner, C., & Patychuk, D. (2010). *The Effect of Standardization in Multicriteria Decision Analysis on Health Policy Outcomes.* Proceedings of the Advances in Intelligent Decision Technologies: Second KES International Symposium IDT 2010: Smart Innovation, Systems and Technologies 4, Baltimore.

Zagana, E., Obeidat, M., Kuells, C., & Udluft, P. (2007). Chloride, hydrochemical and isotope methods of groundwater recharge estimation in eastern Mediterranean areas: a case study in Jordan. *Hydrological Processes, 21*(16), 2112-2123. http://dx.doi.org/10.1002/hyp.6390

Zagona, E. A., Fulp, T. J., Shane, R., Magee, Y., & Goranflo, H. M. (2001). Riverware: A generalized tool for complex reservoir system modeling. *Journal of the American Water Resources Association, 37*(4), 913-929. http://dx.doi.org/10.1111/j.1752-1688.2001.tb05522.x

Zeitoun, M., Allan, T., Al Aulaqi, N., Jabarin, A., & Laamrani, H. (2012). Water demand management in Yemen and Jordan: addressing power and interests. *The Geographical Journal, 178*(1), 54-66. http://dx.doi.org/10.1111/j.1475-4959.2011.00420.x

Zeitoun, M., Messerschmid, C., & Attili, S. (2009). Asymmetric Abstraction and Allocation: The Israeli-Palestinian Water Pumping Record. *Ground Water, 47*(1), 146-160. http://dx.doi.org/10.1111/j.1745-6584.2008.00487.x

Zeleny, M. (1987). Management support systems: Towards integrated knowledge management. *Human systems management., 7*(1), 59-70

Zemann, M., Wolf, L., Hötzl, H., Sawarieh, A., Seder, N., et al. (2012). *Pharmaceuticals as long term reacers of intentional and unintentional wastewater reuse in the Lower Jordan Valley.* - Poster presented at the IWRM 2012, Karlsruhe, November 21-22 2012.

Ziegelmayer, T. (2008). Rational begründete Wasserverteilung : Gerechtigkeit und Nachhaltigkeit im transnationalen Kontext des Jordantals. *Doctoral Thesis*, FernUniversität in Hagen, Hagen.

Zwicky, F. (1969). *Discovery, invention, research: through the morphological approach.* Toronto: Macmillan.

7.   Appendix

A: Structural Analysis of the Jordanian Water Strategy:
"Water for Life: Jordan's Water Strategy 2008-2022"

Institutional setting

The need for an institutional reform is understood as central necessity to support
the accomplishment of many of the other objectives (**Fig. 3.28**).

Strategic objectives	Implementation approaches
Revise, update and extend water legislation	- Develop a new water law to clarify institutional responsibilities in the water sector - Set the NWMP as a binding document in the new water law - Revise existing laws and standards to permit prosecution of water strategy goals - Develop legislation to revise the role of traditional water rights in Jordan's future
Reform institutional structures to clear responsibilities and improved efficiency	- Separate policy-making and planning from bodies responsible for service operation - Separate bulk and retail water provision - Establish institution and sector performance monitoring bodies - Establish management or other contracts with public and private companies - Separation and corporatization of individual water sector utilities - Enable and foster water user groups through training and technical assistance - Enhance collaboration among all concerned ministries
Water sector capacity building	- Establish National Water Training Center for better vocational and continuous on-the-job-education programs for all water sector employees - Upgrade and expand utilities and work environments of water sector institutions

Resource Availability

Assess, monitor and exploit all resources at rates that can be sustained over time

Strategic objectives	Implementation approaches
Assess and monitor available resources	- Conduct periodic assessment of all available and potential resources - Manage treated wastewater as integral part of the national water budget - Improve, automate and centralize monitoring and data collection - Improve analysis and modelling capabilities and ensure its use in decision making
Maximize available renewable resource quantities	- Increase storage by reservoir sediment removal and construction of new reservoirs - Increase groundwater safe yield by artificial recharge (also with treated WW) - **Increase volume of wastewater captured and treated** - Increase volume of brackish water desalination - **Rainwater harvesting in new buildings and industry compounds** - Utilize grey water in new (high-rise) buildings and industry compounds - Regulated allocation (import/export) of resources in accordance to demands - Dig new wells in all governorates - Implement the Red-Sea-Dead-Sea-Canal project
Utilize fossil groundwater	- Formulate a long-term plan for exploitation and revolving five year plans from that - Non-renewable groundwater is only allocated to supply municipal and industrial demands

	-	Implement the Disi Water Conveyance Project
Transboundary Water Resources	-	Pursue bilateral and multi-lateral regional co-operations and contracts for monitoring and usage of shared resources

Resource Protection

Maximize the protection against pollution, quality degradation and depletion of resources.

Strategic objectives	Implementation approaches
Reduce the risk of pollution	- Systemise and automate quality monitoring of surface and groundwater - **Establish protection zones for recharge areas of drinking water supply resources** - Conduct vulnerability and hazard assessments for ground- and surface water - Reduce discharge of untreated wastewater - Enforce quality standards for drinking water, industrial wastewater discharges and treated wastewater - Develop guidelines for pollution risk reduction of potentially hazardous activities - Establish criteria in order to apply the "polluter pays" principle
Reduce risk of depletion	- **Reduce all groundwater withdrawals towards safe yield of aquifers** - Enforce and continuously monitor abstraction limits on private wells - Close water wells that extract water from deteriorating and depleted aquifers - Close all illegal water abstractions
Environmental and health protection	- Conduct Environmental Impact Assessments and Management Plans during planning, design and construction of all water sector development projects - **Consider environmental quality and quantity demands in sensitive habitats when designing water allocation plans** - Limit the use of brackish water in irrigation in order to minimize soil salinity - Monitor observation wells near waste water treatment facilities - Enforce and sustain national health standards in municipal water supply and treated wastewater reuse

Supply Efficiency

Distribute water to consumers in adequate quantity and quality and at the required time to meet the demand in the most efficient manner.

The following allocation principles are stated in the national water strategy:

- First priority is given to the allocation of a modest share of 100 liters per capita per day to domestic water supplies
- Nonrenewable groundwater sources shall be allocated to municipal and industrial uses as a first priority
- Maximize the use of alternative water sources including the use of greywater and rainwater harvesting in industries, municipalities and homes
- Encourage the use of marginal groundwater quality for agricultural uses especially when such use may relieve pumping from fresh groundwater aquifers.

- All treated wastewater generated will be used for activities that demonstrate the highest financial and social return including irrigation and other non-potable uses.
- Use all surplus treated wastewater for irrigation whenever safely possible while ensuring that health standards for farm workers as well as consumers are reinforced.
- Desalination of brackish water in agriculture and industry.

Strategic objectives	Implementation approaches
Provide continuous water supply	- Prepare and continuously update a Balancing and Allocation Plan - Plan water supply under consideration of various future scenarios (including climate change) - **Rehabilitate old and damaged supply sources and network components** - **Expand production and supply networks to required capacities in all regions** - Reduce service response time for connections, complaints and repairs
Ensure adequate supply quality for intended usage **Improve operating efficiency of water allocation**	- Improve monitoring and quality testing facilities for water supply (laboratories) - Apply quality-bound allocation priorities - Use multiple resources mix to maximize the net benefit from the use of unit flow - Reduce physical supply and storage evaporation losses with maintenance program - Improve planning, operation and maintenance, technical, managerial and financial capability of supply departments by human resource and infrastructure programs
Improve cost efficiency of supply service towards the final goal of full cost recovery	- Simplify and eventually eliminate the current supply subsidies - Revise and rise water tariffs, connection and well fees considering differential prices for water based on water quality, the end users and the social and economic impact of prices on economic sectors and regions - Link cost recovery to the average per capita share of the GDP and its level in domestic water and the cost of living and the family basket of consumption - **Improve meter reading and billing accuracy to reduce administrative losses** - Assess future industrial, commercial, tourism and agricultural projects in their requirements of units of water flow - Reduce man-power required per unit of water delivered to customer - Enable and promote private sector participation in water supply services
Improve energy efficiency of the water supply and distribution systems	- Partner with manufacturers of pumping stations and facilities to introduce energy efficiency programs at all working facilities - Establish incentive for renewable energy use and energy efficiency in water sectors - Develop a plan for replacing farming with solar production in the Jordanian Desert - Produce renewable energy from wastewater sludge

Demand Management

Improve water use efficiency and reduce water demand in each sector while creating awareness of water value

Strategic objectives	Implementation approaches
Improve water awareness in every demand sector	- **Conduct national programs for water education and conservation** - Implement water harvesting, reuse and conservation in all official institutions - Provide transparent information on water sector issues to foster understanding of existing problems and necessary reforms

	- Encourage participation of basin water users associations in decision making - Use water tariffs to better represent the value of water consumed
Increase overall water use efficiency	- Provide loans for water harvesting, reuse and conservation technologies - Establish incentive structures for water saving in every sector - Enforce standards for new buildings to reduce water consumption - Promote greywater use, rainwater harvesting and maximum flow limits in domestic facilities - Strengthen Water User Associations with responsibility in water management - Buy private highland-wells from their owners and forbid new private wells
Reduce irrigation water demand and improve irrigation efficiency	- **Reduce the irrigated area in the highlands** - Prevent the expansion of irrigated agricultural areas in the Jordan Valley - Ban summer crops in drought years - Re-allocate freshwater from agriculture to the other sectors - Promote drip irrigation systems - Advise farmers to focus on crops of high-quality and with higher economic returns per unit of water used - Permit regulated water market between agricultural users - Rent lands from farmers in order to prevent them from planting crops

Waste Water Management

Wastewater is collected, treated, managed, and used in an efficient and optimized manner with due consideration to sustainability, economy, quality assurance of the effluent and the requirements of public health and the environment.

Strategic objectives	Implementation approaches
Increase waste water service coverage	- **Priority for full waste water service coverage in resource recharge areas** - **Complete collection in urban areas served by treatment facilities** - **Rehabilitate all sewerage pipes which are over 10 years old** - **Install decentralized treatment to serve semi-urban and rural communities** - **Establish and enforce standards for the use of septic tanks in rural areas** - Treat sewage from un-served areas in well monitored and maintained facilities
Treatment quality meets national effluent standards	- Effluent quality should suffice standards for irrigation reuse and aquifer recharge - Enforce continuous and systematic monitoring - **Upgrade WWTP when necessary** - Enforce standards for industry wastewater discharge to sewers - Provide incentives for industries to treat their wastewater to effluent standards
Put all treated wastewater to productive use	- Create public understanding and acceptance of treated effluent reuse options - Provide decision and implementation support for treated wastewater reuse - **Use treated effluent in agriculture, industry, landscapes and aquifer recharge** - Oblige industries to partly recycle their wastewater - Encourage power generation from sludge - Establish sludge as fertilizer and soil conditioner
Improve cost efficiency of waste-water services	- **Set wastewater charges to cover at least cost of operation and maintenance** - Revise wastewater connection fee - Enable and promote private sector participation in waste water services

A: Software employed for the Dropedia-platform

Component	Author/Developer	License	Reference (all last accessed on: 02/10/2012)
Basic Modules			
Debian	Debian Project	GPL-2.0+	http://www.debian.org/
Apache HTTP Server	Apache Software Foundation	Apache License 2.0	http://httpd.apache.org/
PHP	The PHP Group	PHP License	http://www.php.net/
MySQL	Oracle (formerly Sun, MySQL AB)	GPL-2.0	http://www.mysql.com/
Squid	The Squid project	GPL	http://www.squid-cache.org/
Wiki-Modules			
MediaWiki	Wikimedia Foundation	GPL-2.0+	http://www.mediawiki.org/
Apc	N. Laxström	NN	http://www.mediawiki.org/wiki/Extension:APC
ImageMap	T. Starling	Any OSI-approved	http://www.mediawiki.org/wiki/Extension:ImageMap
ParserFunctions	T. Starling	GPL-2.0	http://www.mediawiki.org/wiki/Extension:ParserFunctions
Maps	J. De Dauw and others	GPL-2.0+	http://mapping.referata.com/wiki/Maps
Semantic MediaWiki	M. Krötzsch, D. Vrandecic, J. De Dauw, Y. Koren & various others	GPL-2.0+	http://www.semantic-mediawiki.org/
Semantic Forms	Y. Koren and others	GPL-2.0	http://www.mediawiki.org/wiki/Extension:Semantic_Forms
Semantic Maps	J. De Dauw and others	GPL-3.0+	http://mapping.referata.com/wiki/Semantic_Maps
Semantic Internal Objects	Y. Koren	GPL-2.0	http://www.mediawiki.org/wiki/Extension:Semantic_Internal_Objects
Semantic Result Formats	Various	GPL-2.0+	http://semantic-mediawiki.org/wiki/Help:Semantic_Result_Formats
Spark	J. De Dauw	GPL-3.0+	http://www.mediawiki.org/wiki/Extension:Spark
Validator	J. De Dauw	GPL-3.0+	http://www.mediawiki.org/wiki/Extension:Validator